Development Trends in Building Services Engineering

Development Trends in Building Services Engineering

Edited by Tin Tai CHOW

City University of Hong Kong Press

First published 2009

ISBN: 978-962-937-162-3

Published by
          City University of Hong Kong Press
          Tat Chee Avenue
          Kowloon, Hong Kong
          Website: www.cityu.edu.hk/upress
          E-mail: upress@cityu.edu.hk

Printed in Hong Kong

Contents

Detailed Chapter Contents

Foreword

The name "building services" (BS) has been used to collectively denote the range of facilities in buildings and other structures that the building services engineers design, install, operate and maintain. The BS systems are important to the comfort, health, and safety of the building occupants, as well as to building sustainability, conservation and environmental protection. For the building industry, it is not only an application of science for human to combat extraneous influences, but also it has to evolve on the inner core value of humanity.

Traditionally, the design, installation and operation & maintenance (O&M) of BS systems are undertaken by different BS practitioners at different stages of the building life cycle. The imperfect performance of the BS practitioners will affect not only the quality of the work that they deliver, but also the quality of the work of other BS practitioners at subsequent stages. What being suffered are then the ultimate performance of the BS systems and their users. It is not uncommon to hear designers complaining that their good designs were not fully realized due to poor installation or O&M, and O&M personnel complaining that they could not make BS plants perform better due to poor design or installation.

To be a mechanical engineer was always in my mind when I was at school, but finally I ended up as an electrical engineer at the start of my engineering career. As the founding Chairman of the Building Services Division of the Hong Kong Institution of Engineers about 20 years ago, I do take that the cross-discipline knowledge and technical know-how are highly important for the final success of the electrical and mechanical installations in building projects. The only sense in which professionalism of the Hong Kong construction industry exists as a credible and homogeneous unit must not be limited by provincial boundaries. Each engineering discipline and building profession has its own part to play. As long as they share the same mission and vision they can get along in harmony.

Although there have been many text books and references that educate or guide the engineering students and practicing engineers on the BS technology, very few of them were written by the professional leaders and researchers of Hong Kong, and therefore with a coverage totally relevant and applicable for the Hong Kong environment and beyond. Their wide spectrum of technical experiences, research insights, visions and expert directives are in need to help the building industry to identify the key issues and give directions for today and tomorrow. It is delightful to see the joint efforts of the BS academic team at the Division of Building Science and Technology of the City University of Hong Kong, together with the fellow BS professional leaders of Hong Kong to furnish this valuable document of development trends, regulations and technical information, including prominent case studies like the AsiaWorld-Expo and the 2008 Olympic National Stadium ("Bird's Nest").

I sincerely recommend this book to all students, engineers and other professionals who are devoting to the BS related fields and those who would like to capture the uniqueness of the Hong Kong practices from the past to the future.

Ir. Peter Y WONG
President
The Hong Kong Institution of Engineers, 2008–09

Foreword

I am pleased to have been asked to write a foreword for *Development Trends in Building Services Engineering* which documents the exciting technological and other developments in the built environment in Hong Kong and beyond. Hong Kong continues to be at the forefront of some of the most remarkable and inspirational advances in architectural design and construction technology driven by its ever increasing and dynamic population. From 1999 to 2005, over 200 skyscrapers were completed in Hong Kong, making it the world city with the largest number of skyscrapers, outstripping New York by some 30%. These buildings, together with the new generation of skyscrapers exceeding 300 metres, of which Hong Kong currently has five, do not build and plan themselves. They need creative, innovative architects, designers, engineers, surveyors, and construction technologists to name a few of the professions involved in planning and developing these modern age mega-structures.

City University of Hong Kong is proud to be a part of the miracle in built environment that epitomises our modern and dynamic city by providing many of the talented and dynamic graduates who play a major part in these feats of planning and construction. The Division of Building Science and Technology at City University trains and educates young people from Hong Kong and around the world, to meet the ever growing challenges of providing us with a built environment which reflects the achievements and aspirations of our species. The Division is at the forefront of teaching and research into new technologies related to a diverse range of products with built environment applications and therefore produces graduates ready to tackle the challenges posed by our climate, increasing demand for green energy, a healthy and efficient environment, new electrical and other automated installations, and the need to house an ever increasing population in modern, environmentally friendly homes.

Increasing globalisation now means that we compete, and are compared with other world cities, not just in terms of the height of our buildings, but also the extent to which we embrace and apply modern technologies in their construction and maintenance. This book is full of innovative ideas and practical solutions which will inspire students of Building Services Engineering, and all those interested in the application of building services technology to provide a safe, healthy and comfortable environment. Even greater challenges face those who work in this field over the next few decades as our global society calls for even more innovative approaches to ensure development is increasingly sustainable for the benefit of our economic, social and cultural prosperity. The contributors to this book give me great confidence that we can meet these and other, as yet unclear challenges, and that the future of our built environment is in good hands.

Professor Paul LAM
Vice-President (Student Affairs)
City University of Hong Kong

Preface

The built environment and building services systems in Hong Kong have undergone substantial changes in the past decades. As a modern city ranked world number one in the number of tall buildings, Hong Kong has lots of experiences in building services technology that can be shared with the world community.

Chapter 1 of this book gives a snap shot of the facts and figures, and the events that have had remarkable influences on our built environment. Also addressed is how these influences have generated new ideas, practices and regulations. With the economic growth and the elevated demands of the community on quality of life and social responsibility, the use of advanced analytical tools for comprehensive building performance becomes essential. Chapter 2 introduces the computer simulation tools that are currently widely applicable in building services designs, such as for building thermal analysis, airflow prediction, and visual/acoustical comfort evaluation. Case studies are used to illustrate how they work in actual practices.

Chapters 3 to 7 cover the key issues and the development trends in various building services installations. Chapter 3 gives a review on thermal comfort and indoor air quality, as well as the various modes of ventilation in use. A brief analysis of the indoor airflow design concepts for modern buildings is presented with the use of computational fluid dynamics tools. The stratum ventilation is introduced as an example of innovative design ideas in progress.

Chapter 4 introduces the importance of energy conservation and energy management, and the development of the building energy codes in Hong Kong. The spirits are illustrated by the applications in central chiller plants. Also introduced is the simulation-optimisation approach which is useful in the design development of large-scale HVAC systems.

Chapter 5 gives a review on the approval procedures, standards, and legislative requirements of fire services installations. The concept of performance-based engineering approach is delivered, together with the examples of demonstrating the application of CFD tool in studying flame and smoke propagation.

Chapter 6 gives an account on "skyscrapers"—a unique feature in the urban development of Hong Kong. Also discussed is how they have imposed challenges on different engineering fields, with an emphasis on electrical technology and installations. Special designs and space demands, like the lift design requirements, power quality and energy loss problems in power distribution are discussed.

Chapter 7 touches on intelligent buildings, their definitions and assessment methods. Also discussed is the IT network that facilitates information sharing and interoperability. This calls for the adoption of open building control and monitoring system approach, and its use to monitor and assess the reliability, safety and efficiency of the electrical supply and distribution system.

Chapters 8 to 11 switch to look into the building services installations from the angle of applications in different types of buildings. Chapter 8 covers the practical building services designs and provisions in high-grade office buildings. The influences of communication, information and transportation (CIT) on human intelligence, work productivity and flexibility are discussed. Then the focus changes from office accommodation, operational economy and performance to the more macro perspective of sustainability.

Chapter 9 gives a flow of the developments in public housing estates in Hong Kong in the past 50 years, in meeting the growing prosperity of the general public and the expectations of quality of public housing. The social responsibility and the changing role of the Housing Authority are addressed by going through the committed working life of a building services engineer in Housing Department.

Chapter 10 turns to the advanced building services provisions in a modern architecture—the AsiaWorld-Expo international exhibition centre. The various special features in supporting all types of events from exhibitions to major arena-style shows are introduced, such as the utilities tunnel network, ice storage system, textile air distribution system, and the like.

Chapter 11 gives illustrations from a consultant firm of Hong Kong on how to tackle sustainable building design through building physics techniques. Two semi-opened space projects that have received local and international Green Building Awards: the Hong Kong Disneyland MTR Station and Beijing Olympic National Stadium are selected as examples to demonstrate how to utilise natural energy sources which are tailored to the local climate.

The materials covered and the line of flow of the book can be useful for the college final year students working on their projects, the engineers looking for a systematic review of the current technology and the development trends, and any others who are interested to know about the built environment of Hong Kong. This is an attempt to gather the wisdom of the professional engineers and academics to document our state-of-the-art. Hopefully, this will encourage more talents to work along this direction.

T. T. CHOW, Editor
December 2008

List of Illustrations and Tables

Figures

Tables

Acronyms and Abbreviations

Development Trends
in
Building Services Engineering

1

The Challenges on the Built Environment of Hong Kong

The social, economic and technological developments in different continents have inspired advancements in architectural design, construction technology as well as building services installations. In the 20th century, the advancements in science and information technology, the awareness of global environmental changes, and the knowledge on illnesses and medical treatments have raised public concerns on better environmental health, comfort, safety, convenience and conservation.

In this chapter, we will give a snap shot of the facts and figures, and the events that have had remarkable influences on the development trends of the built environment in Hong Kong. How these influences have generated new ideas, practices and regulations on the building features and services systems will also be addressed.

Tin Tai CHOW

Building Energy and Environmental Technology Research Unit
Division of Building Science and Technology
College of Science and Engineering
City University of Hong Kong

1   The Building Development in Hong Kong

Hong Kong is a place full of miracle. Beginning as a fishery and trading port in the 19th century, Hong Kong turns into a leading financial centre within 200 years. Geographically, Hong Kong is at the south of the Guangdong province, at the east side of the Pearl River Delta and facing the South China Sea (see Figure 1.1). Belonging to the subtropical climate zone, Hong Kong's weather is characterised by its hot humid summer with occasional showers and thunderstorms, and dry cool winter with the occasional cold front bringing strong chilly winds from the north. After World War II, the increased immigration from the Mainland China brought to its population growth and the low-cost labour supply, and as a result, the growth of the textile and manufacturing industries. As Hong Kong rapidly industrialised, the economy became driven by exports

Figure 1.1 Hong Kong territory with reclaimed land shown in dark colour

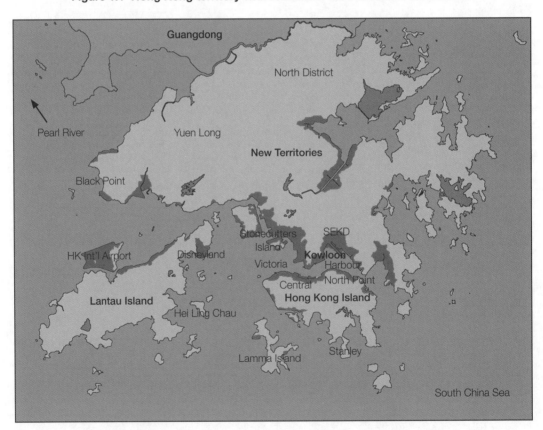

to international markets. Together with Singapore, South Korea and Taiwan, Hong Kong gained the reputation as one of the "Four Asian Dragons" for its high growth rates and rapid industrialisation between the 1960s and the 1990s. Living standards rose steadily with the industrial growth. Although Hong Kong is a small place with a land area of 1,100 km² (including the reclaimed land), its population today is growing to 7 million. The current people density is close to 6,400 people per km² and the household density 2,240 household per km². With the expected influx of immigrants from the Mainland in the multiples of ten thousands per year, these figures continue to grow. About 30% of Hong Kong's population live in public rental housing flats with another 18% in subsidised home ownership flats. These are facilitated by the Housing Authority. The remaining over 50% is with the private housing. Due to the lack of space in the city, few historical buildings remain in Hong Kong. Older buildings are regularly torn down to make way for urban developments. This makes the city more or less a centre for modern architecture. Dense commercial skyscrapers line the coast of Victoria Harbour, a valuable natural asset in the minds of the Hong Kong people no less than the Sydney Harbour in the minds of the Australian. Embraced by the surrounding mountains, Hong Kong's skyline ranks one of the best in the world.

Because of the massive building boom from 1999 to 2005, 225 skyscrapers over 150 metres were completed during the period. At present, Hong Kong has the world's greatest number of skyscrapers—most of these were built in the past two decades—well ahead of the second place New York City by more than 30%. As of 2008, five buildings in Hong Kong are of height exceeding 300 metres. The tallest building is currently the International Finance Centre II which was built in the Central District in 2003 and stands 415 metres tall. One eye-catching development project underway is the 118-storey International Commerce Centre located in West Kowloon. Upon its completion in 2010, this 484-metre new skyscraper will become the tallest building in Hong Kong, and will be ranked the third tallest in the world counting the top occupied level[1].

Like in most other modern cities, the social, economic and technological developments inspire advancements in architectural design, construction technology, as well as building services provisions. The rapid developments in technology and communication, the general awareness of global environmental impacts and the lessons from influenza pandemic have elevated people's expectation for better environmental health, human comfort, life safety, and task convenience. The demands have led to new emerging concepts on engineering design and facilities management. So despite Hong Kong's reputation of being intensely urbanised, the territory has made much effort in the last decade to promote a green environment. As an example, the recent public concern has prompted the severe restriction of further land reclamation from our most famous natural landmark—the Victoria Harbour. There are other successful and unsuccessful stories in relation to infrastructure development and building construction. Presented in this chapter is a snap shot of the events that have influenced the development trends of the building and services system design and operation. Also addressed, here and in the following chapters, are how these influences have generated new ideas and called for new regulations on the building features and services systems. I would like to begin the journey with the hottest environmental concern—the global warming effect.

2 The Global Climate Change

2.1 The IPCC Observations

You may not feel surprised if I mention to you that in 2007, the land temperature in the world reached the historical high level ever since there were scientific records, and was higher than the average value of 1971–2000 by 0.67°C. This global warming phenomenon has been closely watching by the Intergovernmental Panel on Climate Change, which is better known as the "IPCC". The following is a re-organised abstraction from their *Report on Climate Change 2007* [2].

> *"The warming of the climate system is unequivocal, as is evident from observations of increases in global average air and ocean temperatures, widespread melting of snow and ice and rising global average sea level. Observational evidences from all continents and most oceans show that many natural systems are being affected by regional climate changes, particularly temperature increases. The records show, out of the last twelve years (1995–2006), 11 rank among the 12 warmest years in the measured global surface temperature since 1850. The 100-year linear trend (1906–2005) of 0.74°C is larger than the corresponding trend of 0.6°C (1901–2000). The temperature increase is widespread over the globe and is greater at higher northern latitudes. Land regions have warmed faster than the oceans. Rising sea level is consistent with warming. Global average sea level has risen since 1961 at an average rate of 1.8 mm/yr and since 1993 at 3.1 mm/yr, with contributions from thermal expansion, melting glaciers and ice caps, and the polar ice sheets. Of more than 29,000 observational data series (from 75 studies) that show significant change in many physical and biological systems, more than 89% are consistent with the direction of change expected as a response to warming. Changes in atmospheric concentrations of greenhouse gases (GHGs) and aerosols, land cover and solar radiation alter the energy*

balance of the climate system. Global GHG emissions due to human activities have grown since pre-industrial times, with an increase of 70% between 1970 and 2004. Carbon dioxide (CO_2) is the most important anthropogenic GHG. Its annual emissions grew by about 80% between 1970 and 2004. Atmospheric concentrations of CO_2 (379ppm) and methane CH_4 (1774ppb) in 2005 exceed by far the natural range over the last 650,000 years. Global increases in CO_2 concentrations are due primarily to fossil fuel use, with land-use change providing another significant but smaller contribution."

The IPCC *Special Report on Emissions Scenarios*[3] projects an increase of global GHG emissions by 25 to 90% (CO_2-eq) between 2000 and 2030, with fossil fuels maintaining their dominant position in the global energy mix to 2030 and beyond. Even if the concentrations of all GHGs and aerosols had been kept constant at year 2000 levels, a further warming of about 0.1°C per decade would be expected. According to the IPCC *Technical Paper on Climate Change and Water* released on 9 April 2008, there is abundant evidence that freshwater resources are vulnerable and have the potential to be strongly impacted by climate change. Over the 20th century, precipitation has mostly increased over land in high northern latitudes, while decreases have dominated from 10°S to 30°N since the 1970s. For the Asia region, some projected regional impacts are:

(i)   By the 2050s, freshwater availability in Central, South, East and South-East Asia, particularly in large river basins, is projected to decrease;

(ii) Coastal areas, especially heavily populated megadelta regions in South, East and South-East Asia, will be at greatest risk due to increased flooding from the sea and, in some megadeltas, flooding from the rivers;

(iii) Climate change is projected to compound the pressures on natural resources and the environment associated with rapid urbanisation, industrialisation and economic development; and

(iv) Endemic morbidity and mortality due to diarrhoeal disease primarily associated with floods and droughts are expected to rise in East, South and South-East Asia due to projected changes in the hydrological cycle.

2.2 Climate Changes around Us

In the South China Sea region, monsoon affects much the regional climate and the global eco-system. Analysis of the monsoon climate clearly shows the sensitivity of the monsoon system in response to global mean temperature changes. El Nino-Southern Oscillation exhibits the greatest influence on the inter-annual variability of the global climate. Local climate change may follow the global trend. But it may also be affected by the other factors such as urbanisation, irrigation, and desertification. For example, urbanisation leads to an increase in suspended particulates in the atmosphere and thus a decrease in visibility. As a result, the surface solar radiation decreases. The rise in temperature during daytime is then reduced, but this may be more or less offset by the heat released from air conditioning equipments and other urban activities. The net result can be insignificant change in the daily maximum temperature.

By carrying out trend analyses on meteorological observations in Hong Kong, the annual mean temperature data was found to have an average rise of 0.12°C per decade from 1885 to 2002. In the 56-year period after World War II (1947–2002), the annual-average daily minimum temperature T_{min} followed a rising trend of 0.28°C per decade. On the other hand, there was very little change in daily maximum temperature T_{max} but a mild increase of daily mean temperature T_{mean}. The increases in annual-average T_{min} and T_{mean}, as well as the reduction in daily temperature range, can be attributed to the global warming and some other local effects such as the urban heat island (UHI) phenomenon. Comparable occurrences can be found in Macau, though the temperature rise was less significant in this neighbouring city with less population density[4]. In the period 1947–2000, there were long-term increasing trends of T_{mean} and T_{min} respectively at 0.07°C and 0.25°C per decade, but there was no obvious trend in annual-average T_{max}. For the whole 20th century, there was again no asserted linear association of annual average T_{max} with time. T_{mean} was increasing at a mild rate of 0.047°C per decade. All these figures support that the UHI occurrence in the cities raises nighttime temperatures more than daytime temperatures.

2.2.1   What about the future trend in Hong Kong?

In a press release on 12 March 2008, the Director of the Hong Kong Observatory Mr. C. Y. Lam briefed about the updated projection of the temperature trend in Hong Kong in the 21st century[5].

"Taking into account various GHG emission scenarios and the effect of urbanisation in Hong Kong, by the end of this century there will be a temperature increase of 3.0°C for the low-end scenario, 6.8°C for the high-end scenario, and 4.8°C for the middle-of-the-road scenario... In summer, the number of hot nights (with minimum temperature ≥ 28°C) will increase. By the end of the century, the 'middle-of-the-road' number of hot nights is 41 per year. The 'high end' figure is 54. The corresponding figure at the end of the last century was 15. Similarly, the number of very hot days (maximum temperature ≥ 33°C) will also increase. By the end of this century, the 'middle-of-the-road' figure is 15 per year. The 'high end' figure is 19. The corresponding figure at the end of the last century was 7. The average of all calculation results based on different scenarios shows that by 2030–2039,

there will be less than one cold day a year, meaning that for some winters, there will not be any cold days at all... For the situation in which the high emission scenario is coupled with continued urbanisation, the time for this to occur will be advanced to 2020–2029. We will all have the chance to witness the disappearance of winter in Hong Kong."

2.3 International Protocols

Hong Kong emits around 0.2% of the global GHG emissions. In 2005, the emission was 6.5 tonnes per capita. This was far below the level of most developed countries and 6% less than the 1990 emission level. But a continuous drop is not confirmed. Under the terms of the Kyoto Protocol, China (including Hong Kong) is not required to limit its GHG emissions, but has to submit "national communications" to the United Nations. Pressures of changing the requirements have been received in more recent meetings.

On the other hand, Hong Kong is committed to the Montreal Protocol on substances that deplete the ozone layer. The consumption of hydrochlorofluorocarbons (HCFCs) are progressively phasing out. Since January 2004, the import quota of HCFCs has been reduced by 35%. This will drop by a further 65% in 2010 and by 90% in 2015. At the Meeting of Parties to the Montreal Protocol held in September 2007, the members agreed to shorten the phase-out schedule. To cope with the new time frame, Hong Kong needs to further reduce the maximum allowable annual consumption of HCFCs from the original level of 48.6 to 34.7 weighted tonnes by 2010.

The annual consumption of HCFCs in Hong Kong was about 51.1 weighted tonnes in 2007. If this consumption level continues in the coming two years, Hong Kong will have to cut the local consumption by at least 16 weighted tonnes of HCFCs (equivalent to about 300 metric tonnes of R-22) by 2010. As most of the HCFCs consumed locally are for servicing air conditioning and refrigeration systems, the shortage of supply may have adverse impacts on the operation of these systems. The search for and switch to alternatives brooks no delay.

3 The Living Environment in Hong Kong

3.1 The Economy in the Last Two Decades

Hong Kong underwent a rapid transition to a service-based economy in the 1980s with the GDP growth averaged 7.2% annually. Much of the manufacturing operations moved to the Mainland China and the industry constituted only 9% of the economy in this period. As Hong Kong matured to become a financial centre, the growth slowed down to an annual average of 2.7% in 1990s. The economy suffered a 5.3% decline during 1998, in the aftermath of the Asian financial crisis. A period of recovery followed, with the growth rate reaching 10% in 2000, although deflation persisted. From 2001 to 2003, the Hong Kong economy underwent a distinct slowdown which was firstly prompted by the downturn in the US economy suffered by the 911 terrorist attack in 2001, and later the outbreak of SARS locally in 2003 giving a three-year average of 2.0% per year in GDP growth. A revival of foreign and domestic demands led to a strong recovery in 2004, as the cost declines strengthened our export competitiveness. The 68-month deflationary period ended in mid-2004, with the consumer price inflation drifting at near zero levels. The economic growth was robust from 2004 to 2006, giving an annual average of 7.7%. Together with a moderate inflation rate, this is a clear indication of sound economic fundamentals.

Since the old days under the British sovereignty, Hong Kong has been a highly capitalist economy built on a policy of free markets, low taxation and government non-intervention. Despite the ripples, the GDP per capita of Hong Kong currently exceeds those of Japan and many European countries like the United Kingdom, France, Germany, Switzerland and Denmark. We are benefitted by the Mainland's open-door policy and economic reform. As an international finance and trade centre, the city has the greatest concentration of corporate headquarters in Asia Pacific. But the continuous deteriorating air quality in the last two decades has raised the alarm that the corporations may move their headquarters out of our territory. In fact, not only the business community, all parts of the Hong Kong society are worried about air pollution[6].

3.2 Air Pollution

The air pollution problem has been with Hong Kong for some years with no significant improvement despite the efforts spent[7]. Figure 1.2 shows for the period 2000–2007, the percentage of hours within the year that the Air Pollution Index (API) exceeded 100. API, at a scale 0–500, is the conversion of the ambient respirable suspended particulate (RSP), sulphur dioxide (SO_2), carbon monoxide (CO), ozone (O_3) and nitrogen dioxide (NO_2) concentrations measured at the government air quality monitoring network, which is in charge by the Environmental Protection Department (EPD). An index at 100 corresponds to the short-term Air Quality Objectives (AQOs) established under the Air Pollution Control Ordinance. Within the index range of 101–200 at the roadside, those people with existing heart or respiratory illnesses should avoid prolonged stay in areas with heavy traffic. From Figure 1.2 the worst was recorded in 2004 with 7.3% at the roadside that the API > 100, followed by 2007 at 6.7%. According to EPD, the pollutant levels were particularly serious in the Tung Chung and Yuen Long areas, which are

Figure 1.2 Percentage of hourly API exceeding 100 at EPD air quality monitoring network

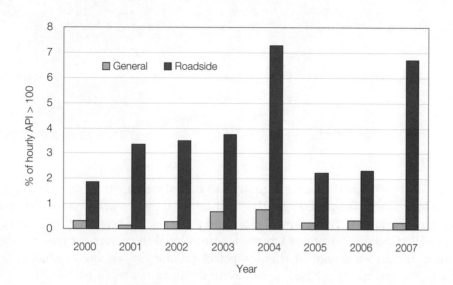

geographically close to the Pearl River Delta industrial area. There are two main sources of air pollutants identified in Hong Kong: (i) the vehicle emissions, and (ii) the GHGs from thermal power plants.

In a survey of vehicle emissions in the periods from May to September 2003 and from November 2003 to February 2004, the pollutant levels were monitored at a number of locations including ambient, roadside and tunnel[8]. It was found that at six out of the seven measuring sites, the concentration of particulate matters at nominal 2.5 µm (PM2.5) was found higher than the United States Environmental Protection Agency Standard of 65 µg/m^3, on 24-hour average basis. In the worst case, the recorded PM2.5 was 288 µg/m^3 inside one vehicle tunnel. And the worst case at the roadside was 129 µg/m^3. The main pollutant source was identified from diesel fuel. PM2.5 assessment corresponds to small suspended particulate matters that may penetrate through the respiratory system of the human body and may reach the lung, and therefore is hazardous. So far Hong Kong has set acceptance level on PM10 but not on PM2.5. Volatile organic compounds (VOCs) are the other key air pollutants. During the survey, the highest record of 797 µg/m^3 was observed at the measuring station outside one Cross-harbour Tunnel.

In fact, the EPD has had a long-running program to control vehicle emissions. In 2000, the requirement of using low-sulphur-content diesel fuel was introduced. In 2005, over 90% of the taxis and one-third of the mini-buses were converted to use less polluting LPG (liquefied petroleum gas). Nevertheless, some of the cross-border vehicles

Table 1.1 Progress in achieving the 2010 emission targets (EPD 2007)

	1997 Emission (tonnes)	2005 Emission (tonnes)	Changes 1997–2005	2010 targets
SO_2	64,500	84,600	+31%	-40%
NO_X	110,000	93,800	-15%	-20%
RSP	11,200	7,200	-36%	-55%
VOC	54,400	40,200	-26%	-55%

are using diesel fuel available at Guangdong and the sulphur content is much higher than those in Hong Kong. From 2006 to 2007, all heavy diesel vehicles had to be installed with approved emission reduction devices and those newly registered had to comply with Euro IV standards. Incentive scheme is also offered to private car owners: the first registration tax for environment-friendly petrol private car has been reduced. On the other hand, the drivers are advised to switch off their engines when waiting.

VOCs are also emitted from a number of seemingly everyday products, such as paints, printing inks and hairsprays. Since 2007, new VOC regulation has been imposed in phases to prescribe VOC limits on selected products. New products in excess of the prescribed limits are banned.

The SO_2 level has increased in Hong Kong since 1997, and the main sources are from the two power companies, the Hong Kong Electric Company Ltd. (HEC) and CLP Power Hong Kong Ltd. (CLP). From 2006 onwards, HEC and CLP have retrofitted their coal-fired units with flue-gas desulphurisation (FGD) systems as an effective means of reducing SO_2. HEC will retrofit two more coal-fired units with FGD and low NO_X burners between 2009 and 2010. CLP will also install selective catalytic reduction systems, which further reduce NO_X, in its four coal-fired units between 2009 and 2011. Another direction is the switch from coal to natural gas for electricity generation. Natural gas combustion emits 90% less SO_2 and RSP than coal, and 80% less NO_X. CLP introduced its first gas-fired generating unit in 1996 and currently operates eight gas-fired units at the Black Point Power Station. HEC commissioned its first gas-fired unit in mid-2006 and converts an existing oil-fired unit to gas firing in 2008. The Hong Kong and China Gas Company also started using natural gas in the town gas production process, and accordingly, a reduction of CO_2 emission from the gas company by 20% was achieved in 2007.

The smog that hangs over the Pearl River Delta is a combination of emissions from Hong Kong and Guangdong. The governments of both jurisdictions have been working together for several years to tackle this problem. The targets, as listed in Table 1.1, have been set to reduce the levels of key pollutants to well below the 1997 levels by 2010. Also listed is the status in 2005, after a progression of work done in Hong Kong. The situations were unfortunately disappointing in 2007 for both parties, making the 2010 targets difficult to achieve.

In October 2006, the WHO (World Health Organization) required the member countries to review their own AQOs and to work out a plan to achieve stringent new targets. One of their concerns is on PM2.5. Hong Kong was among the frontline members in support of the WHO. The work in Hong Kong encompasses an 18-month comprehensive review and the aim is to come up with new standards and long-term management strategy in 2009.

3.3 *Water Resources and Treatment Issues*

3.3.1 Water supply

For several decades after World War II, reservoirs remain the major source of fresh water supply in Hong Kong. Using reservoirs to collect rainwater was found to be grossly inadequate in 1963–64 when restricted water supply to the general public (one time within four days) was once imposed. Thereafter, the then Hong Kong Government (HKG) determined to purchase water from Guangdong. This reliable fresh water supply from the Pearl River has given much support to our urbanisation and economy growth. In 1981, Hong Kong experienced the last water supply restriction. From then on Hong Kong people are enjoying 24-hour water supply at all time. Nevertheless, the use of fresh water for cooling tower application in commercial air-conditioning systems was not allowed before the turn of the century. In 2000, the HKG launched a pilot scheme on the wider use of fresh water for water-cooled air conditioning in non-domestic buildings in the designated areas in Hong Kong. The building owners within the designated areas are thereon encouraged to install water-cooled chillers together with evaporative cooling towers.

In the 1970s, the HKG also imported the desalination technology as an alternative means of fresh water supply. But the rapidly escalated fuel price after the 1973 oil embargo finally led to the termination of the project. Although the utilisation of seawater in this aspect was not that successful, using seawater for toilet flushing as legislated in the 1960s has been highly rewarding. This greatly reduced the demand for fresh water and also enhanced public hygiene. In 1991, about 65% of Hong Kong's households were using free seawater for flushing. By 1999, there were 29 seawater pumping stations along the sea front and seawater constituted 23% of the total consumption of water in Hong Kong. Today, about 80% of the population is provided with seawater flushing.

For decades, seawater was also used for condenser cooling (either directly or indirectly through plate-type heat exchanger, and occasionally for seawater cooling tower application) in central air-conditioning systems, as well as in steam condensers of the thermal power plants for electricity generation in Hong Kong. At this end, Hong Kong has taken the leading role. Nevertheless, seawater quality remains an issue to be dealt with in the 21st century.

3.3.2   Seawater quality

Water quality of the Victoria Harbour was never a problem in the past when the sewage flows were much lighter. For years sewage was dumped untreated into the harbour and flushed away by tidal currents. This approach more or less worked until the early 1980s, when the growth in population and economic activity created sewage loads that were well beyond the holding capacity of the harbour. In 1989 the EPD unveiled a strategic sewage scheme, later known as the Harbour Area Treatment Scheme (HATS), to address this problem[9]. Stage 1 of the project involved the collection and transportation of the sewage from Kowloon and northeast Hong Kong Island to a sewage treatment plant on the Stonecutters Island for chemically enhanced primary treatment. This was commissioned in late 2001 with immediate positive results. Dissolved oxygen, which is essential to marine life, has increased by 10% overall. Compliance with the harbour's Water Quality Objectives was 90% in 2006 as against 50% in 2001. However, not all problems have been solved.

HATS Stage 1 treats 75% of sewage flows around the Victoria Harbour. The remaining 25%, generated by about one million people living in the north and the west side of the Hong Kong Island, are still dumping into the harbour without treatment. Furthermore, the treated effluent from Stonecutters is not disinfected, resulting in an excessive concentration of bacteria in the western waters. Four nearby beaches had to be closed for hygiene reasons, in addition to three others that were closed earlier because of local pollution problems.

HATS Stage 2 is divided into two phases, for making it more manageable and affordable. Stage 2A is to collect the rest of the sewage from the Hong Kong Island and disinfect the treated effluent from the Stonecutters. This is expected to cost HK$8 billion in capital expenses and $420 million in annual operation. It will be commissioned in 2014 provided the proposed sewage charge increases are approved by the legislature. Stage 2B, at the estimated costs of $10.8 billion for construction and $700 million for annual operation, will introduce a higher level of treatment (i.e. biological treatment) and will accommodate future sewage loads. However, several uncertainties may affect the planned schedule, including the future population growth and sewage load, as well as the need to sort out some complex land issues at the site earmarked for its development.

The schedule of Stage 2B will be reviewed in 2010–11, taking into account the latest relevant trends in pollution, load and water quality.

The HATS stage 1 for sewage currently costs about $320 million a year to operate, a figure that will rise to $740 million when Stage 2A is in service in 2014. Currently, only 54% of the applicable sewage services costs are recovered through the sewage charge; the rate has not been changed since 1995. In terms of the share of sewage services costs borne by the customers, the charge is actually declining. This is not only unsustainable, but also unfair because polluters are not paying their share. Therefore, the future direction is expected to increase the sewage charge to reflect the true cost of cleaning up the Victoria Harbour.

3.3.3 Reclaimed water

"Reclaimed water" refers to the sewage effluent that has been highly treated to make it safe for other uses, such as toilet flushing and irrigation. This has been widely practised in developed countries, like the United States, Japan and Australia. In autumn 2006, the EPD commissioned an advanced treatment plant and dedicated pipeline to treat some of the effluent from the Shek Wu Hui Sewage Treatment Works in the North District for producing high quality disinfected water. The reclaimed water serves a selected group of domestic premises, schools and elderly homes for non-potable uses like toilet flushing, landscape irrigation and water features. The demonstration scheme ended in 2008 and hopefully the results will form a basis for the wider use of reclaimed water in Hong Kong. At present, around 1.3 million Hong Kong people still rely on fresh water for toilet flushing, consuming 82 million m^3 of water per year that costs $330 million.

3.4 Solid Waste Treatment

The situation with municipal solid waste is more drastic because households do not pay any direct cost for collecting, treating and disposing waste. This is again inconsistent with the "polluter pays" principle, and also offers no incentive to the community either to reduce or to recycle waste. The three landfills in Hong Kong are running out of space and there are limitations for building new ones. It is therefore imperative to introduce new measures to reduce our waste loads.

The introduction of construction waste disposal charges in January 2006 set an important precedent. Since the charges came into effect, the construction waste disposed at the landfills has dropped by 40%. At this end, the municipal solid waste charges if introduced may also help to reduce waste.

The Chemical Waste Treatment Centre (CWTC) has been operating since 1993 and currently treats around 40,000 tonnes of chemical waste a year. It has been designed to meet stringent environmental standards and has performed admirably, with an environmental and engineering consultancy study concluding in 2006 that the CWTC does not cause any adverse environmental impacts. However, international environmental standards have changed in recent years. The European Union has some of the toughest standards in the world and, while most of the current CWTC's emission levels still meet those standards, they fall short for SO_2 and NO_X. The contract for the CWTC expired in April 2008, providing a good opportunity to upgrade its air pollution control system. The CWTC will also be installed with additional facilities in 2008–09 for receiving and treating clinical waste, thus enabling controls to be implemented on the treatment and disposal of this waste.

Getting people to recycle more can only be effective if there are outlets for their waste. The EPD has been developing a $257 million EcoPark to provide affordable, long-term land for the use of the recycling and environmental industries. The first three lots at EcoPark are for recycling plastics, waste tyres, wood, plaster, glass and other building materials. In addition, 36 short-term tenancies on 7.4 hectares of land elsewhere have been leased to recyclers. The EPD also operates the Kowloon Bay Waste Recycling Centre for the promotion of recovery and recycling electronic products. Since September 2005, the centre has been using by two separate charitable groups for running two pilot programs. A third pilot program has been started in 2006 to recycle cathode ray tubes and recover useful materials from waste computer monitors and televisions. At the end of 2007, the television broadcast of Hong Kong entered the High Definition Digital Era. Accordingly, many families started to replace their televisions with the digital ones. In the first quarter of 2008, the number of cathode ray tubes from waste televisions and computers received at the centre has already exceeded the annual quota. The lack of planning is expected to result in the final handling of these in the private recovery centres. As many of these private centres are located in the North District, the dismantled electronic wastes are easily polluting the soil and the nearby rivers.

3.5 Infection Diseases

Because of the early health education, professional health services, and well-developed health care and medication system, Hong Kong is among the healthiest places in the world. The people here enjoy an average 82-year-long life expectancy, which is the second highest in the world. The infant mortality rate is 2.94, the fourth lowest in the world. Among the world lowest is the fertility rate at 0.95 children per woman; this is far below the 2.1 children per woman required to sustain the current population. However, like many other world class cities, Hong Kong did suffer from influenza infections, which from time to time impose new demands on medical as well as on building services provisions.

3.5.1 Influenza pandemic

An influenza pandemic is a global outbreak of disease that occurs when a new influenza virus appears (or re-emerges) in the human population against which the human has no immunity, then it spreads rapidly (in less than a year) and causes disease worldwide. Pandemics recur periodically, yet unpredictably, and are invariably associated with high morbidity and mortality as well as great social and economic disruption. The three influenza pandemic occurrences in the 20th century were the "Spanish flu" in 1918–19, the "Asian flu" in 1957–58, and the "Hong Kong flu" in 1968–69. The 1918–19 pandemic killed approximately 20–40 million people worldwide, more than the death toll in World War I. And during the last pandemic (H3N2) in 1968, 15% of the Hong Kong population was infected and there were about 33,800 fatal cases outside Hong Kong, mostly in the U.S.

3.5.2 Avian influenza

For more than 10 years, the outbreaks of avian influenza infection among poultry have been reported worldwide. The World Health Organization (WHO) has expressed concern that the avian influenza virus may re-assort its genes with those from a human influenza virus, thereby acquiring the ability to move easily from human to human and thus triggering a pandemic. Avian influenza viruses do not normally infect species other than birds. The first documented infection of humans with an avian influenza virus occurred in Hong Kong in 1997, when the H5N1 strain infected 18 people, resulting in six deaths. To prevent further infection, the HKG health officials slaughtered the chicken and duck populations. Close to two millions were slaughtered. Another outbreak of Influenza A (H9N2) occurred in 1999 with two people infected and thereafter, one case in 2003, and one case in 2007. To prepare for large-scale outbreaks, the public healthcare system, staff and hospital bed mobilisation plans have been prepared. Air-conditioning provisions in the hospitals have been reviewed.

3.5.3 Severe acute respiratory syndrome

In late 2002, an outbreak of severe acute respiratory syndrome (SARS) began in Guangdong. In February 2003, a doctor who had treated cases in Guangdong checked into a hotel in Hong Kong and infected up to twelve other guests there. A large number of healthcare workers were infected while treating patients in different hospitals. It was then from Hong Kong the virus carried to other places and subsequently, posed an enormous threat to the international community. On 30 March, the HKG authorities quarantined the Block E of the Amoy Gardens, a private housing estate, due to a massive outbreak with more than 200 cases at the same block. The virus was brought into the estate by an infected visitor who was discharged from the hospital. Most of the cases were tied to apartment units at the same orientation and shared the same sewage pipe. One speculation of the virus spread was through excretion, and therefore through the wet drainage system. The other speculated theory in support of airborne transmission was the spreading through dried up U-shaped P-traps in the drain pipes, and was blown by a maritime breeze to the apartments via stairwells.

For weeks, there were many infected cases in Hong Kong and not until 24 May that the number of newly infected patients dropped to zero for the first time. Within the period, there were 1,755 identified SARS cases (299 of whom died) in Hong Kong and more than 400 infected patients were healthcare workers. The outbreak has heightened the public concerns on the possible dire consequences of building neglect and the perennial environmental hygiene problems. After the crisis, the review of the building services installations was not limited to hospital facilities but also the drainage system at large. In response, the Buildings Department has conducted a territory-wide survey to all private buildings on defective drains and participated in improving environmental conditions of some identified black spots.

3.6 Noise Impact

The city of Hong Kong is unusual and challenging for noise control. Activities go on till midnight, and sometimes overnight. In many old urban areas, residential, commercial and industrial premises are mixed together without clear demarcation. The background community noise is substantial since their developments were well before proper environmental and planning guidelines were in place. Many people are living next to noisy restaurants, bars, industrial processes and other potentially irritating activities. Sometimes it is not possible to eliminate the problem, but only to reduce its magnitude. The disclosure of noise information in properties sales brochures is a growing community expectation. As a matter of fact, the noise environment often changes. There could be nearby construction work, short-term amusement parks, or new railway line or road. Currently, noise from general construction works between 7 p.m. and 7 a.m. and on public holidays is controlled through construction noise permits. Noise from industrial or commercial activities is controlled by means of noise abatement notices, in that the one emitting the excessive noise is required to reduce it within a given period.

Comparing with stationary noise sources, traffic noise is perhaps the more difficult subject. In a densely populated city, heavy vehicles often have to travel near homes, hospitals, schools and many other sensitive receivers. As a result, more than 1.1 million people in Hong Kong are affected by excessive traffic noise of above 70 dBA. Traffic noise is expected to worsen over the next ten years due to increased traffic, especially in the early morning hours and at night. A study was commissioned by EPD in late 2006 to review the practices overseas in respect of noise standards and to keep abreast of international developments. Also reviewed are the Professional Practice Notes on Road Traffic Noise, which are used by architects, engineers and town planners in designing buildings and new roads, to provide more protection against traffic noise to the residents of new buildings.

4 The Integrative Efforts

A clean and healthy environment is core to our quality of life and the competitive edge of Hong Kong. The success requires the cooperation among the government, the public utilities, the private sector and the users as well. All parties are committed to providing high quality and sustainable living environment.

Sustainable development refers to one that meets the needs of the present without compromising the ability of future generations to meet their own needs. As a matter of fact, sustainable development is a co-operative concept. It encompasses not only environmental protection and conservation, but also holistic thinking and integrated approaches in balancing social, economic, environmental, and resource needs. A number of non-profit organisations, such as the Professional Green Building Council (PGBC) and the Hong Kong Building Environmental Assessment Method (HK-BEAM), have been instrumental in the growth of sustainable buildings. A comprehensive environmental assessment of buildings has been adopted, i.e. from design, construction to management of green buildings.

The Action Blue Sky Campaign, launched by the HKG in July 2006, is a wide-ranging publicity campaign to raise public awareness about energy conservation in all sectors of the community. The campaign encourages the community to take actions in their daily lives to improve air quality, such as setting air-conditioned room temperatures at 25.5°C in summer months. The Chief Executive, Sir Donald Tsang, set an example by launching a "Dress Down in Summer" campaign to encourage civil servants to dress lightly in summer to engender a more comfortable work environment under the 25.5°C initiative.

4.1 Sustainable Urban Planning

4.1.1 Screen-like buildings

The real estate developments in Hong Kong basically follow the urban street pattern. Single blocks of building are packed along streets and most of them are managed independently. The quality then varies from block to block. In the last 30 years, the concept of private housing estate emerged and each estate is self-sustained, having its own shopping mall, sports club and sometimes even schools. Integrative and structured housing management is provided. The idea is particularly welcome by the middle class from the "convenience and economy of scale" points of view. Within a single estate, the number of tower blocks can be from less than ten to over a hundred. With the release of building height restriction after the relocation of the international airport out from the city centre, the new blocks can be from 30- to 70- storey. In particular, private developers are maximising revenues by constructing uniform blocks on seafront sites to give all apartment units some unrestricted sea view. Some buildings, like those in North Point, are actually hiding the beautiful skyline of Hong Kong.

Another controversy is on the "wall effect" caused by uniform high-rise developments, known as the "screen-like buildings". The wind blockage not only adversely affect air circulation and intensify the urban heat island effect, but also impact public hygiene and contribute to air pollution. A survey conducted by a green group in 2007 found that among the 138 private residential developments completed in recent years, around 75% could be classified as screen-like buildings[10]. Moreover, the forthcoming projects linked with railway stations all include high-density screen-like buildings of 50-storey or above.

Back to year 2005, the Planning Department completed a feasibility study on air ventilation assessment (i.e. the AVA Study). A set of design guidelines for the improvement of air ventilation was formulated based on the findings. The AVA Study has proposed a performance-based assessment system to compare the air ventilation impacts of various design options. In July 2006, the Housing, Planning and Lands Bureau and the Environment, Transport and Works Bureau jointly issued a technical circular specifically on air ventilation assessment, making this one important consideration in the planning of major government (re-)development projects. How the new private developments can minimise wind blockage is a matter of urgency in urban planning.

4.1.2 Land reclamation and district cooling

Land reclamation has been a government policy with long history, of expanding the urban areas along the seafront, both within and outside the Victoria Harbour. One recent example was the Hong Kong Chek Lap Kok International Airport, which was constructed in 1990s by reclaiming two small islands at the north of the Lantau Island to serve as the base for the 1,255-hectare airport platform. This was one of the largest earth-moving and dredging operations ever undertaken. A fringe development on this man-made island is a world-class exhibition hall—the AsiaWorld Expo. The relocation of the airport from the city centre to the suburb in 1998 has left a large piece of flat land available for urban redevelopment, known as the South East Kowloon Development (SEKD). This is regarded a favourable site for the application of district cooling technology[11].

District cooling technology is a sustainable means of cooling energy generation through mass production. In 2000, the HKG commissioned a consultancy on "territory-wide implementation of water-cooled air-conditioning systems in Hong Kong", in that three different schemes: cooling tower (CT), central seawater (CS), and district cooling (DC), were compared. It was concluded that the DC scheme is superior to the CS scheme with respect to energy efficiency, building floor space utilisation, and project life cycle cost. Hence the CS scheme should only be considered under special circumstances. Moreover, DC and CT schemes could co-exist in the same district to provide building owners and operators with more choices so as to introduce healthy competition.

The method of condenser heat rejection has a major implication on the central plant chiller efficiency. Generally speaking, seawater cooling could result in the highest overall DC plant efficiency. The continuous discharge of cooling seawater to the territory's coastal waters may elevate the temperature of seawater near to the cooling discharge outfall, bringing thermal stress to the marine ecosystem. This impact can be minimised through source control.

The HKG proposed land reclamation at the Victoria Harbour, as opposed by the green groups, has affected the SEKD (South East Kowloon Development) project implementation. According to the court decision, there should not be any reclamation unless the "public overriding need test" is satisfied, i.e. there are immediate needs and with cogent and convincing reasons; there exists no reasonable alternative; and the proposed reclamation has been kept to minimum.

4.1.3 Environmental impact assessment

The environmental impact assessment (EIA) process, which is backed by legislation, requires proponents of major projects to identify and mitigate environmental problems before final approval can be obtained. EIA must be carried out as a part of the engineering feasibility study of urban (re-)development projects with an affected area of more than 20 hectares or involving a population of more than 100,000 people. Since 1998, the Environmental Impact Assessment Ordinance has protected more than 1.5 million people and more than 1,000 hectares of ecologically sensitive areas from adverse environmental impacts by requiring proponents to incorporate prevention and mitigation measures. Up to early 2008 more than 100 EIA reports have been approved. The process, from the earliest planning stages to project completion, involves a high degree of public inputs and transparency. Hong Kong Disneyland is an example of a major project that underwent an EIA from the earliest planning stages to completion.

In recent years, the scope of EIA has been expanded to encompass strategic environmental assessments (SEAs). The driving force behind SEAs is to promote environmentally sustainable policies. The emphasis is on the impact of policies, strategies or plans such as major land use planning studies, rather than projects.

4.1.4   Renewable energy

In 2000, a study commissioned by the Electrical and Mechanical Services Department (EMSD) found that solar power, energy from waste and wind energy have the potential for wider use in Hong Kong. In 2005, the HKG set a target of generating one to two per cent of Hong Kong's total electricity supply from renewable sources by 2012. Accordingly both power companies have been exploring the application of wind power in Hong Kong. In February 2006, the HEC commenced operation of its 800 kW wind turbine at Lamma Island for public demonstration and technical evaluation, while CLP has planned for the operation of its first wind turbine at Hei Ling Chau in 2008. In addition, the two power companies conducted EIA studies for building off-shore commercial wind farms in Hong Kong waters. The HKG is also promoting the adoption of renewable energy in its discussions with the power companies over the post 2008-regulatory regime for Hong Kong's electricity market, such as providing financial incentives in the form of a relatively higher return for renewable energy infrastructure.

In Hong Kong, the primary use of solar energy at present is to provide hot water for swimming pools and for the slaughterhouse in Sheung Shui. Some small-scale photovoltaic and wind systems have been installed in remote areas to generate nominal electrical power for lighting and on-site data recording equipment. To assist the public to better understand the technical issues and the application procedures relating to grid connection of small-scale renewable energy installations, the "Technical Guidelines on Grid Connection of Small-scale Renewable Energy Power Systems" was made available to the public in 2005. This was followed by a revised edition completed in December 2007 which extends the applicable capacity limit of the original guidelines from 200 kW to 1 MW.

4.2 Public Health and Safety

4.2.1 Fire safety

Fires in old buildings without the provision of modern fire safety measures can be disastrous. There are about 1,400 commercial, 9,000 composite, 3,000 residential and 1,700 industrial buildings in Hong Kong which were constructed with pre-1987 fire services (FS) standards, i.e. with much deviation from the current FS Code requirements. During fire incidents in tall buildings, most fire fighting and rescue operations have to rely on the provisions within the building. The risk is therefore high.

On 20 November 1996, a tragic fire outbreak at the Garley Building, an old 15-storey commercial building in South Kowloon, resulted in 40 fatalities and 81 injuries. The Discovery Channel labelled this over 20-hour catastrophic event as the "Hong Kong Inferno". On that day lift replacement was underway with the lift cars of two lifts being removed and their landing doors at all floors being opened. Bamboo scaffolding was erected inside the lift shafts where electric arc welding work was carried out on Level 15. Many smoke detectors on Level 2 were disabled by wrapping them up with plastic bags in order to stop unwanted alarm. The welding sparks fell through the lift shaft and caused ignition at the Level 2 landing. Unfortunately smoke and burning smell from the ongoing welding work lowered the fire alert of the occupants. Although the fire occurred during office hours, no one seemed to be aware at the early stage. The lift shaft acted like a chimney so the flame and smoke quickly reaching Level 13 and started another fire there. Such a tragedy could be avoided if proper maintenance procedures and fire precaution measures were in place, and strictly followed by the staff members of the property management and the lift company. Over the years, even for post-1987 buildings, there were many occurrences of fire owing to the temporary close down of fire protection systems and the removal of fire dampers for the convenience of building renovation works.

For years, FS provisions are for dealing with accidental fires, like the one at the Garley Building. This is so not until the recent release of BS 7974, in which the precautions on arson fire (i.e. fire set up purposely by igniting sources) and attack fire (i.e. due to military action or terrorist attack) are also considered. On the other hand, the adoption of the fire engineering approach is recommended. The fire engineering approach requires the designer to look into FS designs from the first principles. As with other engineering disciplines, FS design should allow the incorporation of new ideas and also value engineering. The extent of FS provisions should cope with the nature of the fire risk so that a reasonable level of safety can be achieved, whereas unnecessary intrusion into building design or over expenditure can be avoided. FS systems are the integral parts of the overall fire safety design and should be considered in a holistic manner. Through the fire engineering approach, the performance, reliability and cost effectiveness of FS systems can be readily analysed and improved. For example, water storage requirements, the spacing of fire detectors and sprinkler heads can be determined through our professional knowledge of fire science instead of dictating the code requirements.

4.2.2   Gas safety

Gas explosions may occur inside process equipment or pipes, in furnished buildings or construction sites, in open process areas or confined spaces. It is most important in risk management to lower the explosion probability to an acceptable level, or otherwise this may lead to disastrous consequences.

On 11 April 2006, a fatal underground explosion occurred at the ground level of an old building in East Kowloon, due to the leakage of town gas from underground pipes. Two people were killed and eight injured. The building was damaged seriously. The gas system was managed by the Hong Kong and China Gas Company Ltd. that supplies town gas to 85% of the Hong Kong households, and also to commercial and industrial customers. After the accident, a comprehensive survey was carried out by the gas company to identify any possible accumulation of escaped gas of any kind in voids close to buildings. Openings found were sealed to stop the gas entry or exit, and the data were recorded in the electronic system for future reference. The precautionary work should continue.

4.2.3   Electrical safety

In Hong Kong, the transmission and distribution power cables laid underground are more than 22,400 km in total length. In most urban areas, the pavements are crowded with power cables together with other utility facilities. Damage to a live electricity cable, particularly in the course of excavation, may lead to an explosion. While the site workers and the passersby may be electrocuted or burnt, the damage may also cause power interruption and bring serious inconvenience to the general public. The EMSD is currently working together with the power companies to monitor statistically the incidents involving damages to underground cables. In collaboration with trade organisations and other government departments, the Code of Practice on Working near Electricity Supply Lines has been revised, aiming at providing practical and up-to-date guidance for the construction industry.

The code of practices for electrical installations in Hong Kong are derived from BS 7671 (i.e. IEE Wiring Regulations) which is the national standard for the safety of electrical installations in U.K. In addition, there have been modifications in technical requirements to suit the Hong Kong situation and trade practices. To enhance the safety of the public and building owners, communal electrical installations are required to be inspected, tested and certified every five years. On the other hand Hong Kong does not set our own standards on the safety of household electrical products. To further improve the situation, since 2005 the EMSD and the trade had mutually agreed to set a time frame for the supply of electrical products in compliance with the latest IEC safety standards.

From 1 July 2007 onward, cable colour for fixed electrical installations in Hong Kong has been changed from the current red/yellow/blue/black colour to the new color scheme of brown/black/grey/blue, in order to align with the latest international standard. The change has stabilised the cable supply to Hong Kong and minimised price fluctuation.

4.3 Architectural, Services Design and Construction

4.3.1 Green engineering

The HKG carries a leading role in green engineering. Since 1995, legislative control over the Overall Thermal Transmittance Value (OTTV) has been introduced for regulating the building envelope performance of new commercial buildings, including hotels. Later on, a set of comprehensive Building Energy Codes (BECs) that address energy efficiency requirements on building services installations was released. To promote the compliance with BECs, the voluntary Hong Kong Energy Efficiency Registration Scheme for Buildings was also launched. The current effort is to make the BECs mandatory.

In 2006, the EMSD introduced an assessment tool to appraise the life cycle performance of commercial building developments, and to assess their environmental and financial impacts. The tool assists in the selection of alternative materials and/or systems. Many green building features will be brought into the Central Government Complex at Tamar, to be completed by 2011. Supported by the two power companies, the EMSD also periodically organises the "Hong Kong Energy Efficiency Awards" energy saving competition which is targeted at two private sectors: (i) Commercial and Residential Buildings, and (ii) Schools.

The professional institutions also play their part. The Professional Green Building Council (PGBC) continues to launch the Green Building Award, which aims to promote sustainable and green development, to recognise developments and research projects with outstanding contributions to sustainability and the environment, and also to encourage the industry towards wider adoption of sustainable practices in planning, design, construction, maintenance, and renovation projects. The PGBC comprises five local professional institute members, namely, Hong Kong Institute of Architects, Hong

Kong Institution of Engineers, Hong Kong Institute of Landscape Architects, Hong Kong Institute of Planners, and Hong Kong Institute of Surveyors.

4.3.2 Historic building conservation

Many historic buildings in Hong Kong are under the threat of demolition. The Murray House, a Victorian-era building constructed in 1844, now in Stanley Bay, was originally in the Central District where the Bank of China Tower now stands. It was dismantled in 1982 to give way for the modern landmark. The restoration of this building (originally as barracks for the British military) in 1998–99 was a fortune since the preference on modern architecture and infrastructure construction have torn down many other historic monuments without the chance of relocation, like the old Repulse Bay Hotel (1920) and the Central Post Office Building (1911). After all, relocation can only be seen as the last solution as this fails to reflect the historical development and architectural significance as it was in the old days. The existing legislation concerning heritage preservation in Hong Kong has been commented as "piecemeal" and weak.

In 1992 the United Nations Educational, Scientific and Cultural Organization (or UNESCO) launched the "Memory of the World" program to guard against collective amnesia. This program set out to preserve valuable archive holdings and library collections all over the world, and ensure their wide dissemination. Building conservation requires particular expertise and care because historic monuments are significant and invaluable heritage of our culture that once lost or damaged cannot be replaced. The principle is that historic building has a special message from its creator and thus, its original structure and appearance must not be altered or falsified[12]. Development and preservation are not mutually exclusive. From the sustainable development point of view, we do not own heritage, but merely keep them for our future generations. The HKG needs to play a more proactive role, especially when the owner of a private heritage is expected to receive considerable benefits from the property developer through the selling.

4.3.3 Artificial intelligence and telecommunication

In 2004, Hong Kong Cyberport received the internationally acclaimed Intelligent Building of the Year Award 2004 from Intelligent Community Forum presented in New York. This information technology (IT) flagship of Hong Kong was being praised, not just for its state-of-the-art IT infrastructure and beautiful architecture, but more importantly the role model that it sets for future IT-connected community—broadband and information systems technology has been adding demonstrable value in the form of advanced services and merits to its tenants.

Back to 1970s, building automation existed in its early form as single functional control and monitoring of discrete devices or systems. As the devices or subsystems in buildings getting more complex and inter-dependent, multifunctional control of subsystems became desirable for providing an efficient and effective integrated control system. Subsequently different integrated systems flourished in the early 1990s, and the developments had transcended to Building Management Systems (BMS) and Integrated Communication Systems (ICS) in the late 1990s. The more recent development has been to combine these two forming the Computer Integrated Buildings (CIB), which covers not only the conventional building subsystems integration, but also the integration to communication infrastructure to provide a new dimension of value services, and even extended to home automation. The joy of living will rise to a new horizon, when people can wishfully access their homes and control their appliances at some remote sites such as offices through the internet.

As a matter of facts, Hong Kong holds one of the most sophisticated and successful telecommunication markets in the world. This has been an important factor in Hong Kong's development as a leading business and financial centre. Broadband Internet access services are very popular in Hong Kong. Apart from fixed carriers, in January 2008 there were 171 internet service providers licensed to provide broadband services. As at the end of 2007, there were more than 1.87 million registered customers, using broadband services with speed up to 1,000 Mbps (Megabits per second). In the residential market, 76% of the households are using broadband service. Internationally, Hong Kong's broadband penetration rate is among the highest in the world.

4.4 Facilities Management

In the past, operation and maintenance service was never regarded as glamourous work. That may be true. Nowadays, buildings are huge and complex in function, and the building services installations are increasingly complicated. A diverse team with a serious professional attitude is required to take care of the day-to-day operation for achieving high quality and optimum performance. While the designers and the project managers live with the building services systems for some months during the design, construction and commissioning phases, the occupants and the operation and maintenance staff have to stay with them for many years. Post occupancy evaluations and full records are essential to ensure a new building be able to operate in its designed conditions.

On the other hand, as a result of the prevailing weak building care culture, there are increasing number of old and dilapidated buildings in Hong Kong showing signs of urban decay. Proper building management and timely maintenance of existing facilities help to prolong the overall life span of buildings, optimise the economic value of scarce land and improve the living environment.

4.4.1 Indoor air quality

Other than public health, an improved indoor air quality (IAQ) environment bears economic benefits such as reaching higher worker productivity, lower absentees, and lower medical costs. In Hong Kong, the EPD engaged a consultant to carry out a thorough study of the IAQ conditions in offices and public places in 1995–96. This was through questionnaire survey and field measurements[13]. The survey with more than 2,000 respondents found that 32% of the occupants were not satisfied with the IAQ situation in their workplace. This result was comparable to other similar studies done by the WHO. The measurements found that over an 8-hour period, about 38% of the offices were having CO_2 level above 1,000 ppm, and 20% having bacteria level above 1,000 cfu/m^3. The causes of these excessive concentrations were mainly inadequate ventilation and excessive occupancy density. One third of the offices were found to have too much formaldehyde in summer. Actually in many commercial premises, the building management purposely reduced the ventilating rate in order to save electricity costs.

Since 2003, the EPD has introduced a voluntary registration scheme with two classes, i.e. the "excellent class" based on a set of IAQ objectives with 12 parameters, and the "good class". If the IAQ objectives are complied with, an IAQ certificate will be issued for putting up at a prominent location of the building for public information. After the registration, the building owner is responsible for the post-certification IAQ control. As at April 2008, there were 37 excellent class certified premises and 73 good class certified premises on the list.

4.4.2 Energy management

Energy management works with an interrelationship between three microscopic factors: finance, technology and human behaviour, under the impact of four macroscopic factors: physical, economic, social and political climates. The ultimate goal is to pay as little as possible for the energy consumed. Sometimes this may have nothing to do with energy saving, for example, to make use of the power company tariff structure for overnight thermal storage. The importance of human factor is often overlooked during the process. In fact, the best way to achieve efficient use of energy and conserve resources is to have everyone committed and taking action.

Energy audit involves periodic examination of energy use in premises. It becomes popular in both the public and private sectors to check whether energy is being consumed effectively and efficiently. It allows a comparison of energy use in similar periods or similar occasions, leading to the formulation of annual budgets and energy targets for future plan. To help into these, the EMSD has developed a set of energy consumption indicators and benchmarks for specific groups of buildings in Hong Kong. A benchmarking tool is available to allow users and operators to compare their energy consumption levels with others in the same group, set future targets and identify measures to reduce energy consumption. Through energy audits, the HKG was able to reduce its electricity consumption by 5.6% between 2002–03 and 2005–06, which was equivalent to 120 GWh electricity saving and a reduction of CO_2 emission by 84,000 tonnes annually. Further reduction is deemed possible.

4.4.3 Waste recycling

One of the great challenges for Hong Kong's "high-rise" population is finding the space to separate waste for recovery. Most homes do not have room to store recyclable or re-usable materials. The "Program on Source Separation of Domestic Waste", introduced by the EPD in January 2005, aims to provide places close to home where residents can separate their waste and store the recyclable or re-usable materials until they are picked up by recyclers. The types of recyclables collected have been expanded from paper, aluminium cans and plastic bottles to include plastic shopping bags, compact discs and metal tins. Up to mid 2008, more than 940 estates and buildings, representing half of the Hong Kong population, enrolled in the program. About 30% of these estates implemented floor-based waste separation but most estates set up waste separation facilities on the ground-floor. The goal is to have 80% of the population, in more than 1,300 housing estates, participating in the program by 2010. Nevertheless, whether a program is successful or not should not be judged by the number of enrolled estates but the number of families that have actually put this into action. According to statistical findings, only less than 20% of the inhabitants in the enrolled estates are participating in the waste separation program. There are much rooms for improvement.

5   What's in Need?

Quality buildings are essential ingredients of a world class city. In Hong Kong, from time to time our built environment faces various domestic and international influences and challenges. While in the past some of these were global matters, many were coming from our own culture and perhaps, from our less than desirable living habits. We are expected to face greater challenges in the 21st century that will call for more changes of our practices for the benefits of economic prosperity whilst meeting our growing social and environmental aspirations. Sustainable development cannot be preceded just by setting slogans or through demonstration projects. It requires a radical change in the values and attitudes of the community in favour of our future generations.

The application of building services technology in providing safety, health and comfort, is a task which requires continuous and shared efforts of all parties. The government, through setting standards and promoting community participation, is responsible for raising public awareness and exercising reforms on one hand, and regulating the side-tracked behaviour on the other. The public utilities are to provide stable and economical energy sources to the customers with greatest convenience and without introducing environmental pollution and life hazards. The learnt societies and pressure groups are expected to react from their precise observations and visions through the public media, and equally important, to undertake educational and professional development activities. Academics are to search for innovations, evolutions and in depth analyses in response to technological advancements and environmental changes, and at the same time, to re-organise knowledge for the public and the young. For the designers and project managers, it is their roles to alias with the suppliers and contractors to ensure that new installations are delivered in time and in full compliance with the professional standards and statutory requirements. They are also to work in hand with the operation and maintenance team to ensure that performance specifications are fully met, and systems are economically operated and well maintained throughout the life cycle. For the owners, users and general public, it is their utmost safety, health and comfort that moves the technology forward. The basic needs and conservation desire are no doubt the motivation behind the efforts of all parties. With the collective efforts from each and every of us, we together are going to create a marvellous and green built environment for Hong Kong.

Notes

1. Wikipedia, the free encyclopedia. List of tallest buildings and structures in the world, http://en.wikipedia.org/wiki/List_of_tallest_buildings_and_structures_in_the_world.

2. IPCC. (2008). *Climate change 2007, the fourth assessment report of the Intergovernmental Panel on climate change, Vol. 1–3*. Cambridge University Press.

3. Nakicenovic, N. and Swart, R. (Eds.) (2000). *Special report on emissions scenarios*. IPCC.

4. Chow, T. T., Chan, A. L. S., Fong, K. F., & Lin, Z. (2006). Some perceptions on typical weather year—From the observations of Hong Kong and Macau. *Solar Energy, 80*(4), 459–467.

5. HK Observatory. Director of Hong Kong Observatory talks on climate change and 2008 outlook (12 March 2008). www.hko.gov.hk/wxinfo/news/2008/pre0312e.htm.

6. Civic Exchange. (December 2008) Summary of Key Findings, Hong Kong's Silent Epidemic—Public Opinion Survey on Air Pollution, *Environment and Public Health—2008*.

7. EPD. (2007). *Environment Hong Kong 2007, Annual Report*. Environmental Protection Department, HKSAR.

8. EPD. (February 2005). *Final report of determination of suspended particulate and VOC emission profiles for vehicular sources in Hong Kong*.

9. EPD. An overview on water quality and controlling water pollution in Hong Kong. www.epd.gov.hk/epd/english/environmentinhk/water/water_maincontent.html.

10. GovHK. Press release: *Screen-like buildings*, February 28, 2007. www.info.gov.hk/gia/general/200702/28/P200702280173.htm.

11. Chow, T. T., Au, W. H., Yau, R., Cheng, V., Chan, A., & Fong, K. F. (2004). Applying district cooling technology in Hong Kong. *Applied Energy, 79*(3), 275–289.

12. Chui, H. M., & Tsoi, T. M. (August 2003). *Heritage preservation: Hong Kong and overseas experiences*. Report. Faculty of Social Science, University of Hong Kong.

13. EPD. (1995). *Consultancy study on indoor air pollution in offices and public places in Hong Kong*. Final Report. CE 14/95.

2

Computer Simulation— The Indispensable Tool in Modern Design

Worldwide energy and environmental concerns have prompted the local construction industry and the Hong Kong SAR Government to pay closer attention to the sustainability of buildings and to put forward new codes and regulations in governing building performance respectively. These new developments in the industry progressively raise the demands on the use of advanced simulation tools for comprehensive building performance analyses.

We will introduce the state-of-the-art computer simulation tools that are widely applicable to building services designs. Cases will be provided to illustrate how they work in actual practices.

Tin Tai CHOW and Apple Lok Shun CHAN

Building Energy and Environmental Technology Research Unit
Division of Building Science and Technology
College of Science and Engineering
City University of Hong Kong

1 Introduction

1.1 Modernisation of Building Design

In Hong Kong, building design has undergone huge changes. New buildings are bigger, taller, more extensive in functions, and penetrating deeper underground. In commercial buildings, large glazing, huge atrium, and the use of information technology are symbols of modern design. The property developers favour building complex that includes a mix of building types like shopping centre, office tower, hotel and residential block, within the same development project. In the last 30 years or so, the global energy crisis and environmental issues have made conservation of energy a key factor in architectural design. With the awareness of sick building syndrome since 1980s, indoor air quality has become another equally important design factor. Obviously, a building with contemporary aesthetic appearance is far from modern if it is not sustainable, and when the quality of building services is not matching. In taller and larger buildings, there are increasing difficulties in creating comfortable, safe and healthy, yet cost effective artificial environment.

Under the global influences, the Hong Kong SAR Government in recent years has progressively put forward new codes of practice and regulations on energy/environmental performance of buildings and the services installations. Some are mandatory but most are still voluntary. There are increasing efforts on introducing new concepts in architectural features and building services provisions. There is also a trend to go for performance-based system designs, such as smoke extraction performance and building energy budget determination. The stakeholders are calling for the construction of buildings that are equipped with "high-tech" intelligent features, and meeting specific environmental assessment standards. There are demands on wider applications of renewable energy and better promotion of green engineering. All these give much opportunity to the use of advanced predictive tools for comprehensive performance analyses.

1.2 Advanced Predictive Tools

Building performance analyses rely on the tactful use of laboratory tests and/or computer simulation in one way or the other. Physical model studies, such as wind tunnel, environmental chamber, artificial sky dome and the like, are popular in the context of both practical design and academic research. Wind tunnel is very effective in studying wind effects, such as its impact on building structure, the wind pressure at openings in relation to ventilation design, and the dispersion of pollutants over an urban district.

Environmental chamber is a useful tool in the study of indoor air movement and human comfort. Solarscope or artificial sky allows the assessment of the effect of overshadowing and the control of sunlight penetration. The limitations of these laboratory approaches are that they are relatively expensive and the time of experiment can be lengthy. Laboratory results are also subject to experimental errors that are inherent in physical modelling and measurement techniques. Computer analysis, on the contrary is more economical and today, more readily to match the building design and construction schedule when prompt decisions are to be made.

The advancement in computer technology has been amazing since the first computer came into picture in 1950s. The never-ending increase in computing speed and storage, plus the corresponding drop in price make possible the development of advanced numerical and graphical tools to handle complex analytical and presentation tasks. Many public and commercial software packages with sophisticated user interface are now available. Extensive model validation and comparative studies have been reported; some tests were performed by reputable international organisations with participating software developers from around the world. The fast growing use of simulation software has gradually gained the trust of the construction industry towards greater use of simulation technology. Practitioners today generally accept that modelling-and-simulation is a practical approach to assess building performance. It is not exaggerating to say—computer simulation has become an indispensable tool in modern design.

Design analysis supported by computer simulation involves the "creation" of a behavioural model of a building, or a plant system, or both at a given project stage. The simulation task involves executing this model on a digital computer, and analyzing its observable states, which are made up of the post-processed outputs of the simulation runs. Behavioural models are to be developed by reducing real world physical entities and phenomena to an idealised form and at a desirable level of abstraction. From this abstraction, a mathematical model is constructed by applying the conservation laws of physics. At this end, the range of mathematical models derived may span from simple/empirical to comprehensive/fundamental types. The same happens to the simulation software, of which the quality depends very much on the sophistication of the simulation core (or engine) in use.

2 Building Performance Simulation

2.1 Building Thermal Analysis

Heat and mass transfer processes in air conditioning systems remove unwanted heat and moisture from the indoor space. Computer simulation techniques can provide accurate estimation of hourly space heat gains and cooling loads, with which the engineers can carry out optimum design of the heating, ventilation & air-conditioning (HVAC) system, including proper equipment sizing. On the other hand, architects can conduct parametric studies on different building features for screening energy-efficient design options. As a matter of fact, building envelope and environmental services provisions can be seen as an integrated object in thermal performance evaluation. The American Society of Heating, Refrigerating and Air-Conditioning Engineers, Inc. (ASHRAE) published an Energy Standard for Buildings except New Low-rise Residential Buildings, ASHRAE 90.1, with the 2007 edition as the most updated version[1]. The standard provides a performance-based Building Energy Cost Budget Method on building services and building envelope design. Trade-off among the energy performances of various building services systems and building envelope are allowed so that innovative building services and building designs become technically feasible. Computer simulation can be conducted to determine the annual energy cost budget of a newly designed building, in that its compliance with the performance-based energy code can be assessed. The performance-base Building Energy Code of Hong Kong has been developed following the similar spirit.

Another type of building thermal analysis is applied to simulate the transient performance of thermal energy system such as refrigeration plant or HVAC system. Here component-based simulation software is used to evaluate the optimum system configuration and control strategy of a thermal plant, like central chiller plant or renewable energy system. Reduced time step down to seconds can be used, and sometimes even a fraction of a second as in the refrigeration system simulation.

Hence, in principle, there are two main categories of dynamic simulation software for building thermal analysis: (i) the building energy simulation programmes and (ii) the building thermal systems simulation programmes, though the boundary between these two gradually becomes less clear as the energy analysis software continues to develop over the years.

2.1.1   Building energy simulation programmes

Building thermal load varies with time due to natural and activity changes, both indoor and outdoor, and the heat capacity of the building structure that gives rise to thermal effects on a time-dependent manner. Building elements such as external walls, roofs and floor slabs (that work as "capacitors") play an important role in delaying the effects of heat flow. It is necessary to have precise estimation of hourly heat gains through building envelope as well as cooling loads on the unsteady state basis. Heat Balance

(HB) equations and Radiant Time Series (RTS) are two commonly used approaches for the precise estimation[2].

There are a number of building energy simulation software packages available in the market, and those most widely used include DOE-2 and EnergyPlus. DOE-2 employs room weighting factors for the calculation of thermal loads and room air temperatures[3]. This programme has been developed by the United States Department of Energy for years and the current version is DOE-2.1E. Parametric studies can be conducted with different combinations of window-to-wall (WWR) ratios, façade orientations, shading coefficients of glazing, aspect ratios, thermal transmittances of opaque walls and roofs, and many others. A database of built-in air-side systems and central air-conditioning equipment are available in DOE-2.1E subprogrammes. Users can select and incorporate these components for modelling a building. With these input data and information, 8,760 hourly electric and hot/chilled water demand of air-handling system, energy consumption of the central plant as well as the whole building energy consumption, can be evaluated. For new buildings, these results can be used to assess the compliance of performance-based energy code while energy target (in terms of kWh/m^2) for existing buildings can be established for energy monitoring. Any substantial deviation of the measured building energy consumption from the energy target will give the building owner or operator an "alert signal". Action will be taken to check whether the deviation in energy consumption is abnormal or acceptable, for instance, this sometimes is just owing to an unrealistic input of building operation schedule.

In 2001, the US Department of Energy launched a new-generation building energy simulation programme called EnergyPlus. It has included most of the best features from DOE-2 along with new capabilities, such as surface temperature and moisture adsorption/desorption prediction, sub-hourly time steps, and many others. An important provision is a modular structure that facilitates the adding of features and links to other programmes such as those used for daylight and HVAC simulations. Other examples of new algorithms and models developed in EnergyPlus include shadow algorithm and EcoRoof model[4]. In assessing the heat gains due to solar radiation in buildings of irregular shape, it is essential to determine the extent of shading on each external wall surface within a particular hour. A shadow algorithm in EnergyPlus is able to compute such sunlit and shaded portion of a building facade based on coordinate transformation methods. For studying the thermal performance of a building with green roof, an EcoRoof model developed in EnergyPlus can be used; this allows the user to specify various aspects of green roof construction including growing media depth and thermal properties. Moreover, the inclusion of fluid/air loops and user-configurable HVAC systems makes EnergyPlus becoming a comprehensive building thermal systems simulation programme.

2.1.2 Building thermal systems simulation programmes

These programmes, also known as system component-based simulation, are provided with a system component library. They could fulfill the tasks of building energy simulation programmes mentioned above. In addition, they could provide HVAC system-only simulation, or integrated building-and-system simulation. The user is able to interconnect the various components to form a thermal and/or fluid flow network of specific desirable topology. The simulation engine then calls the system components based on the data of the input files and solves the system equations by iteration until a convergent solution is obtained. TRNSYS[5] and ESP-r[6] programmes belong to this category. To suit individual simulation task, users are allowed to construct their own system component models such as specific types of air-handling unit, water pump, chiller, or storage tank, with user defined physical properties including dimensions and heat transfer coefficients.

The modular feature and capability of this type of simulation programme allows users to conduct transient simulations for solar thermal systems, fuel cells, modern renewable energy systems including photovoltaic and wind power under different sorts of external excitation. In building services engineering, various HVAC components and systems can be modelled and simulated. For example, in TRNSYS various heat rejection systems like air-cooled condenser and cooling tower using potable water or marine water can be modelled and their energy performance as well as running cost can be evaluated. With the aid of life cycle costing technique, best option can then be chosen for the retrofit of an existing HVAC system.

While ESP-r here is grouped under the component-based simulation programme category, it is widely used in Europe for building energy performance assessment and hence its application is similar to EnergyPlus. The latest version of ESP-r is 11.5, released in May 2008. One feature which is unique to ESP-r and EnergyPlus is the mass flow network solver. The user is allowed to define the fluid flow networks for air and/or water paths with the user defined pressure heads (like fan curve or pump curve) and the adjustable flow resistances of various types (like automatic volume control damper). Hence the programme is able to determine the mechanical-driven or buoyant-induced fluid flow when pressure nodes of a flow circuit are given or can be determined at each time step.

2.2 *Weather Data for Computer Simulations*

The need of typical weather data for evaluating building energy performance is obvious. In 1970s, several groups of researchers in the United States, including the National

Climate Data Center (NCDC) developed the TRY (Test Reference Year) file, which was meant for a yearly record of 8,760 hourly weather data sets. The selection of TRY from a number of actual weather years required the sequential screening of those years having months with extreme temperatures, until one year finally remained. By experience, this selection process would result in a reference year with weather patterns being too mild for performance analysis purpose. NCDC together with the Sandia National Laboratory (SNL) then created a new data set named as TMY (Typical Meteorological Year)[7]. The revised procedure was to identify the single real months of hourly data, called as the TMMs (Typical Meteorological Months), which could fill up the 12 individual months of the representing year. In the mid 1990s, ASHRAE acquired from NCDC the weather data of 900 international locations outside North America. A project was launched in 1997 to use these data sets for generating International Weather Year for Energy Calculations (IWEC) files[8]. The IWEC files of 227 international sites were finally produced—Macau was on the list but not Hong Kong.

By comparing the use of multiple-year measured weather data against several typical weather data sets, Crawley[9] reported the result analysis of a prototype office building using the DOE-2.1E hourly energy simulation programme as follows: (i) the use of TRY should be avoided since no single-year data could match well with the typical long-term weather pattern; and (ii) the use of a synthetic year, like TMY or IWEC, could be advantageous for its better prediction of the energy consumption pattern as well as the energy costs.

Figure 2.1 Comparison of Hong Kong TMY and TRY solar data with long term average

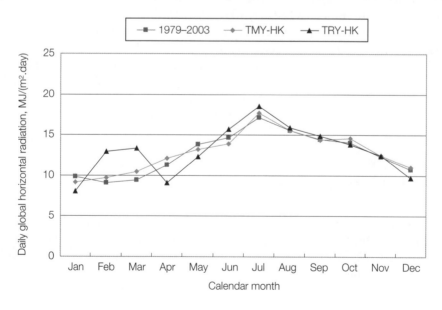

Table 2.1 Comparison of the TMMs of Hong Kong and Macau

Month	Jan	Feb	Mar	Apr	May	Jun	Jul	Aug	Sep	Oct	Nov	Dec
H. K.	1995	1988	2003	1986	**1997**	**1990**	2000	2002	**1982**	1984	1989	1993
Macau	1991	1995	1982	1993	**1997**	**1990**	1991	1982	**1982**	1986	1990	1986

In our process of generating the TMY file for Hong Kong in 2004, the hourly weather data of the 25 year period (1979–2003) available from the Hong Kong Observatory was used[10]. Table 2.1 lists the TMMs of Hong Kong (from our results) and Macau (from the IWEC file) for ready comparison. It can be seen that only the TMMs of May, June and September (bolded in the Table) are the same. This indicates that the occurrences of the most typical monthly weather conditions of the two neighbouring cities were mostly not falling in the same period. Incidentally, the annual solar radiation received in Macau was constantly higher than in Hong Kong on record.

In 1990s, the year 1989 was identified as the "TRY" of Hong Kong, and the weather data set has been specified for use in the performance-based building energy computations. But from Table 2.1, it can be seen that only November 1989 appears as one of the TMMs for Hong Kong, not the other months in 1989. Figure 2.1 shows the comparisons of the monthly average solar radiation (H) for TMY, TRY and the co-incident 25-year mean values in Hong Kong. On monthly bases, the absolute differences of TMY and TRY with the long-term mean values are respectively 0.47 and 1.37 MJ/(m^2.day) on average. On yearly bases, the annual average of H for TMY, TRY and the long-term mean values are respectively 12.9, 13.1 and 12.7 MJ/(m^2.day). So TMY gives better representation than the TRY, and this also happens to other weather indices. On the other hand, TMY file should be periodically reviewed when considering the long-term climatologic change as expected in the 21st century. The TMY weather files of Hong Kong as well as a number of other international cities are available at the EnergyPlus website of the US Department of Energy. Translation programme is available to convert the EnergyPlus weather input file into the ESP-r weather format.

2.3 Air Flow Prediction

Air temperature, humidity, velocity and pollutant concentrations are the key parameters for assessing indoor thermal comfort and air quality. Recommended design values of these parameters are readily available from design standards and energy codes, so the HVAC system can be selected accordingly. However, even when these recommended values are generally satisfied in an indoor space, local thermal discomfort and poor air quality may still exist in specific locations. The problem can be the result of improper air distribution.

The patterns of indoor airflow and pollutant distribution can be predicted by solving a set of partial differential equations for the conservation of mass, momentum, energy and species concentration. Computational Fluid Dynamics (CFD) is a numerical technique applied for this purpose, based on known and/or assumed boundary conditions. A major advantage of using computer analysis over wind tunnel on fluid flow is that it provides results for the entire flow field, rather than just at points on a physical model where pressure tappings are placed. It is particularly useful in areas such as sensitivity analyses of unconventional buildings where there is little prior experience, and for cases where the complete information about the whole flow field is important.

CFD is extensively used both as a design tool and for research purposes. It helps to generate better and faster designs by testing different options. Using the new generation desktop computers, one can run most commercial CFD software. General-purpose commercial CFD packages are available and those popularly used by the research community including CFX, FLUENT and AIRPACK. Almost all methodologies in building simulation use finite volume techniques for the CFD calculation, though finite element is the alternative choice and is more suitable for complicated geometry. There are 3 main groups of airflow problems in buildings: calculation of the external flow around buildings, calculation of the bulk internal flow, and the transient flow simulation. The 2-equation k-ε model with Boussinesq approximation is by far the most popular technique being adopted.

The problem with turbulence modelling is largely owing to the wide variation of length scales. In order to describe the physical processes adequately a high grid resolution is required. This consumes large amounts of computing power and memory. In addition, the large range of time scales further complicates the problem. To overcome this problem, various solutions have been proposed including the RANS (Reynolds Averaged Navier Stokes) solution which gives the time averaged values. These solutions are based on the well known eddy viscosity models. Large eddy simulation (LES) is another type of approximation where the large scales are simulated directly, and the smaller scales are computed based on simple turbulence model. For fire propagation and smoke dispersion in a confined space, the use of the LES approach becomes an advantage, which is adopted in the public-domain software—the Fire Dynamic Simulator.

2.4 Visual/Acoustical Comfort Evaluation

General lighting criteria for visual comfort includes appropriate task illuminance, relative brightness (contrast) and avoidance of glare. Artificial lighting provides reliable illumination for indoor space regardless of the outdoor daylight condition. In some buildings, daylight integrated with artificial lighting has been utilised so that electricity consumption in electric lighting can be conserved. Natural lighting is induced into indoor space where sunlit environment can be created. However, glare and increased air-conditioning load may result. Accordingly, detailed studies are often conducted in the design stage to investigate the visual comfort condition and the impact of electricity consumption in buildings. Lighting performance simulation software can fulfill the need for predicting interior illuminance in complex building spaces due to daylight and/or electric lighting system under a variety of sky conditions.

On the other hand, the study of intruding noise and sound quality is increasingly important. Places like conference room, concert hall and TV studio, should be designed and constructed with optimum reverberant time and free of acoustical defects such as echo, sound shadow and long delayed reflection. Machine plant rooms, in particular the emergency diesel generator room, should be acoustically insulated. In the past, time-consuming and expensive scale models were used to evaluate the acoustic design of concert halls. Nowadays, with the advanced computing technology, computer simulation programmes using ray-tracing technique enable design engineers to predict and assess the acoustic behaviour of any enclosed or open space with various building materials and under various operating conditions.

Lighting performance simulation software provides an effective means of predicting illumination condition for complex building space with electric lighting and/or daylight. Well-known lighting performance simulation programmes include SUPERLITE and RADIANCE. The former is a classic daylighting programme using the flux transfer method as its simulation approach. It is capable of calculating daylight factors and daylighting levels on any given plane for detailed room geometries, standard daylighting techniques and shading from external and internal obstructions. RADIANCE uses the ray-tracing technique to calculate the luminance inside a space of any shape which can be furnished. Moreover, it can create a three-dimensional (3D) and colour rendered visual representation of the space and furniture with shading and calculated luminosity values. A Window version known as "Desktop RADIANCE" that integrates Radiance with AutoCAD is available.

An integrated lighting simulation tool called ADELINE (Advanced Day- and Electric Lighting Integrated New Environment) allows the conversion of CAD drawing of a space to its input file. It links to the above mentioned lighting performance simulation programmes (i.e. SUPERLITE and RADIANCE). While the former calculates illumination levels and daylight factor on any plane, the latter predicts and presents 3D displays of various lighting scenarios in colour, and with complex graphic representations to analyse the luminance distribution, glare sources and visual comfort.

In addition, output results from lighting performance simulation programmes can be used as input data of the building thermal performance programmes such as ESP-r and EnergyPlus. This serves the purpose of studying the energy impact of daylighting on buildings and HVAC systems. To quote an example: a room with utilisation of daylight can be divided into a number of zones and the simulated lighting level for each zone is compared to a design value, say 500 lux for office. Where a zone is over lit, stepped lighting or continuous dimming control can be applied and functions could be generated to describe the relationship between light output and electrical input into the lighting system. Therefore, 8,760 hourly fractions of electric lighting that is lit by daylight for the entire zones are calculated and transferred to building thermal performance programme as input file. The data enables the programme to determine the reduction in electricity consumption of electric lighting and the corresponding increase in space cooling load. Hence the electricity consumption of the HVAC system can be calculated. With these simulated results, design engineers can have a clear picture to judge whether daylight should be utilised in a building for achieving energy conservation.

Four more examples of applying the advanced simulation software in environmental services design in Hong Kong are elaborated in the following section.

3 Application Examples

3.1 Comparative Studies in Building Envelope Performance

In an energy-efficient building, heat gain/loss through building envelope should be minimised. Key design parameters involve building orientation, insulation, window glazing types, window-to-wall ratio (WWR) and building aspect ratio. Building energy simulation programme is a useful tool in the comparative study of building envelope performance.

A generic curtain-wall office building of 40-storey in Hong Kong, with the perspective view of one floor level shown in Figure 2.2, is used to explore the relative thermal performance of the building envelope. In the base case, 0.5 is used for the shading coefficient (SC) of window glazing (single-pane absorptive type), 40% for WWR, whereas no thermal insulation is applied on the external wall. Table 2.2 lists the base case design parameters (bolded), together with other parameters for comparative studies. The public domain software EnergyPlus was used in the study to evaluate the heating and cooling loads of this generic building with various design parameters in separate simulation runs. The simulated annual heating and cooling loads are shown in Figure 2.3, for ready comparison with the base design case. As seen from Figure 2.3(a), the increase in heating load with glazing of SC equal to 0.1 and 0.3 are 32% and 14% respectively, while there is a 11% reduction in heating load for glazing with SC = 0.7. This is mainly due to the reduction in solar heat gain through window with glazing of lower SC in winter season. For cases with larger WWR, despite of the increase in solar heat gain through larger window glazing area, there are 17% and 34% increase in heating loads because the larger window area has lowered the thermal resistance of the whole exterior wall. Hence more heat loss by convection-conduction pair results. Thermal insulation is crucial in the reduction of heat loss through opaque walls. There are 30% and 42% decrease in heating loads for cases with 25mm and 50mm insulation respectively.

Figure 2.3(b) shows the trend of increase in cooling load with increases of SC and WWR. Higher value of SC and larger window area (both having significant effect on the

Figure 2.2 Perspective view of the office building

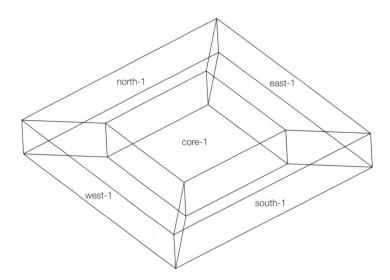

Table 2.2 Key parameters of building envelope in the comparative study

Parameters		
Shading Coefficient	Window-to-Wall Ratio	Insulation in opaque wall
0.1	20%	**No insulation**
0.3	**40%**	25mm insulation
0.5	60%	50mm insulation
0.7	80%	—

Figure 2.3 Comparison of annual thermal loads of the generic building with different shading coefficients, WWR and thermal insulation: (a) heating, (b) cooling.

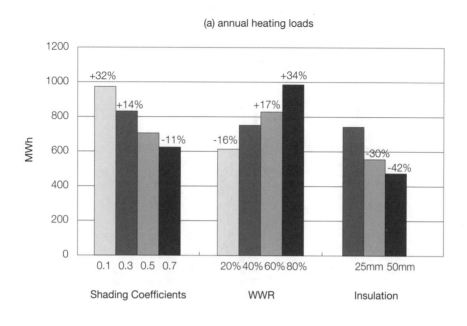

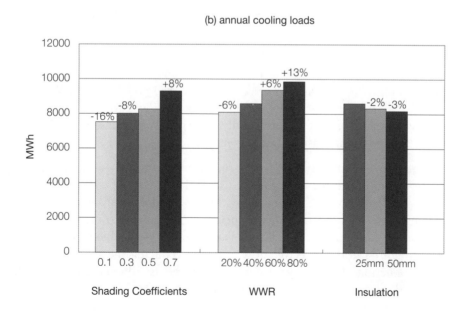

Figure 2.4 Energy flow path at a natural-ventilated double-glazing

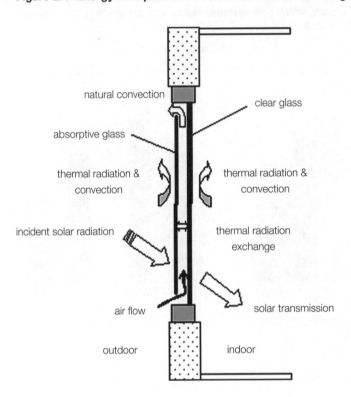

increase in solar heat gain) were found to contribute much to the increase in the cooling load of the building. The increases are 8% (for SC = 0.7) and 13% (for WWR = 80%) respectively. For thermal insulation inside opaque wall, it was found that the reduction in cooling load is not significant. There is only 2% and 3% decrease in cooling loads for cases with 25mm and 50mm insulations respectively.

While the single absorptive glazing is most commonly used in Hong Kong, the green engineering approach may ask for the use of (i) PV glazing (clear float glass panes sandwiched with "transparent" solar cells), and (ii) natural ventilated double glazing. The PV-glazing converts a portion of the incident radiation into electricity, and hence reducing the solar transmission. The latter, as in Figure 2.4, is able to remove the solar heat away through natural convective flow and thus reducing the summer cooling load. The accurate prediction of the buoyant flow requires the use of mass flow network solver and thus the ESP-r software can be used in the analysis. Figure 2.5 shows the relative performance of these two green options compared to the single clear float glazing as the benchmark.

Figure 2.5 Variation of monthly cooling load reduction in percentage relative to clear float window glass

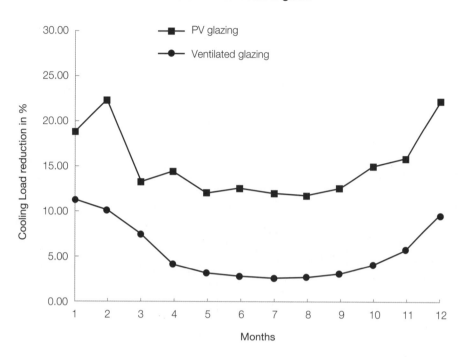

3.2 Energy Performance of a Facade-integrated Hybrid Solar System

Hong Kong has a good potential for solar energy application. In the city, new apartment buildings can reach 70 floor levels or higher, though 30 levels are common. If the orientations of the un-shaded external facades are favourable, these high-rise residences are ideal sites for applying the facade-integrated photovoltaic and water heating (PV/T) technology. The year-round intensity of solar radiation on a vertical surface varies with its orientation. In high-latitude countries, the south-facing wall will receive most radiation. This may not be the case for low-latitude places because of the relatively high solar altitude in summer. In Hong Kong (at 22.3°N), using the TRNSYS programme based on the TMY data set, it is found that the Southwest quadrant receives the highest year-round solar intensity[11].

Component based plant simulation study can be applied to evaluate complex electrical and mechanical system performance. Figure 2.6 shows the schematic diagram of the proposed PV/T system. The two water tanks, for pre-heat and main storage, are

Figure 2.6 Schematic diagram of photovoltaic-water heating system

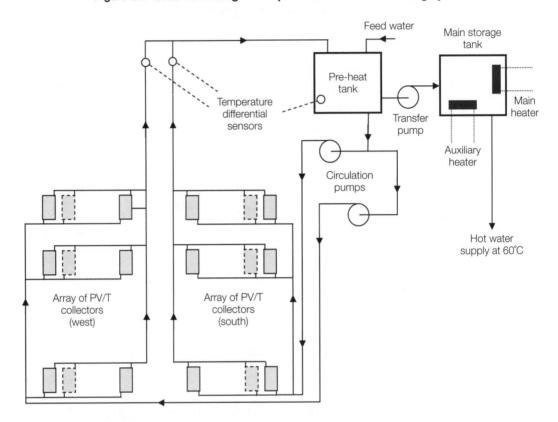

interconnected by a transfer pump. They are located on the flat roof, together with two circulation pumps serving two groups of PV/T collectors on the south- and west-facing facades. When the pumps operate, all solar collectors will receive cooling water at the same flow rate. Water after its heat pick-up at the solar collectors is stored in the pre-heat tank. The circulation pump operating-hour within a day is from 6 a.m. to 7 p.m., which can be overridden by the on-off signal from the individual temperature sensors that monitor the tank supply and return water temperature differential. By priority, electricity generated at the PV modules is first stored in the batteries. Maximum power point trackers and DC/AC inverters are provided. The electrical energy is used to operate the circulation pumps, and to augment the local grid for providing auxiliary electrical heating at the main storage tank. From the pre-heat tank, water is fed to the main storage tank at a rate that is commensurate with the instantaneous hot water demand. The two heaters in the main tank keep the water storage and supply temperature stable at 60°C.

A numerical simulation model that analyzes its energy performance at an apartment building in Hong Kong is first to be developed. It is estimated that with the use of amorphous-silicon hybrid collectors which cover two-third of the west- and south-facing façades, the system is able to support one third of the thermal energy required for water heating[12]. Figure 2.7 shows a simulation flow chart of the PV/T system. The solar radiation model transforms the hourly data of direct and diffuse solar radiation on horizontal surface into the irradiance falling on the vertical surfaces of given orientation and angle of tilt. The make-up water flow rate is determined by a hot water consumption

Figure 2.7 Simulation flow chart of photovoltaic-water heating system

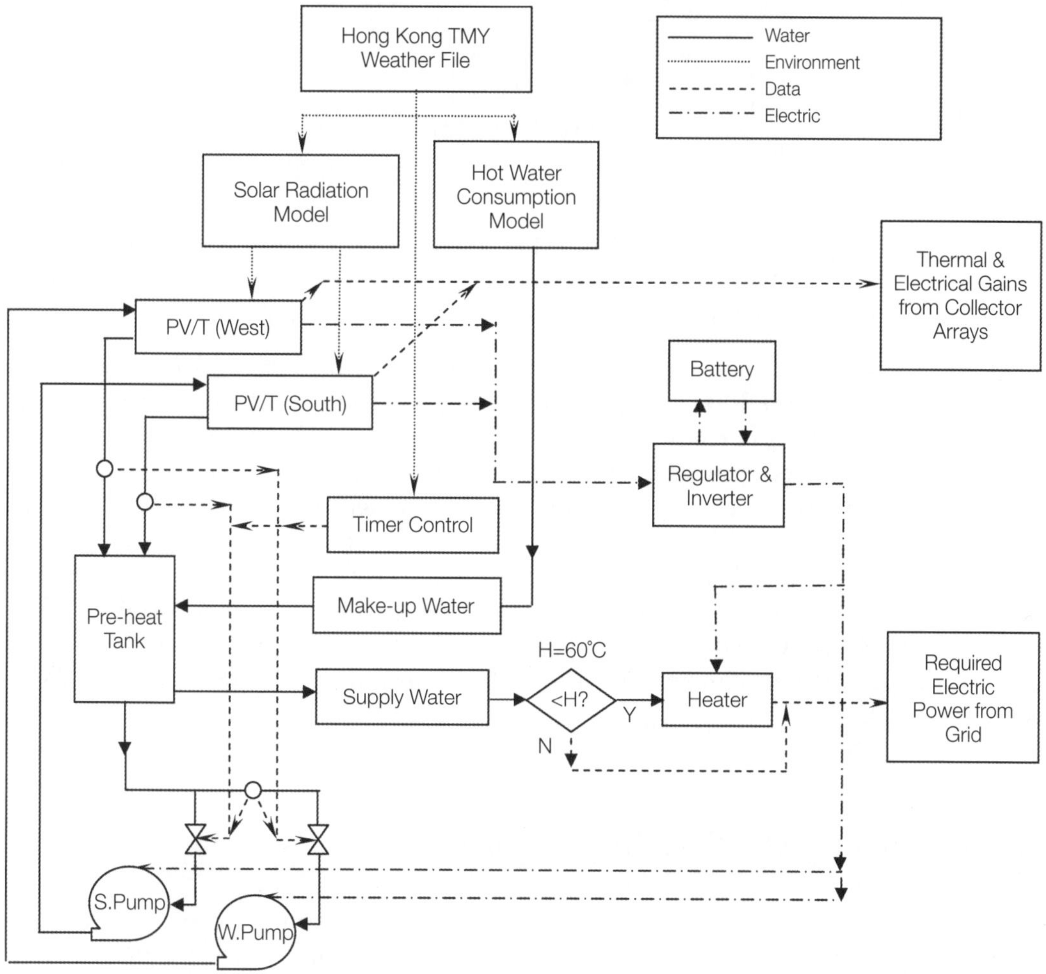

Figure 2.8 Variation of annual energy gain with water flow rate

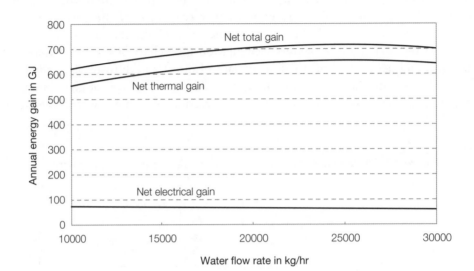

model. This computation process has taken into account the effectiveness of heat conversion as well as losses at various sub-systems. The priority of the PV-generated electric power is set to feed the batteries via the regulator-and-inverter.

In this system, the higher the water flow rate, the better is the cooling effect and therefore, the better the PV/T collector performance. However, the circulation pump power increases with water flow rate. There is an optimal point where the net energy gain (thermal plus electrical gains at collectors less the pump power and the system losses) is the maximum. Through a systematic adjustment of the water flow rate in different simulation runs, the optimal flow rate here is found to be 23,500 kg/hr, as indicated in Figure 2.8. With this optimal flow, the water temperature in the pre-heat tank throughout the year is found generally within the range of 15 to 40°C. The simulation results show that the S wall has the peak gain during winter, whereas the W wall has the peak gain in summer. This shows the compensation effect of using two PV/T collector groups.

3.3 Airflow at Building Re-entrant

Residential building projects in Hong Kong are typically comprised of a number of high-rise residential towers standing on a common podium. The pattern of tower arrangements generally varies depending on the prevailing climate and topography. The widespread development based on a "cruciform" plan for the individual towers is essentially a product of the Building (Planning) Regulations. Figure 2.9 shows an example. Public space including lift lobbies and corridors is usually squeezed to a minimum at the building core. With its four wings radiating out from the core, a residential tower consists usually of eight apartments on a single floor, with two at each wing. A narrow but deep re-entrant, exists at each wing between the two neighbouring flats. The windows of kitchens and bathrooms that are open to a re-entrant are meant for dual purposes of natural lighting and ventilation. Because of the restrictive air movement at the re-entrant, the oily kitchen exhausts are frequently trapped within this space under adverse wind conditions. This results in substantial re-circulating flow back to the kitchens via opened windows. In new residential buildings, a growing trend is to provide split-type air-conditioners to serve the living/dining rooms and the bedrooms. From an aesthetic point of view, the recessed building re-entrants are particularly well suited for accommodating the outdoor condensing units. However, the thermal energy dissipated from the stacks of the condensing units tends to produce a buoyant plume

Figure 2.9 Typical floor plan of a residential tower

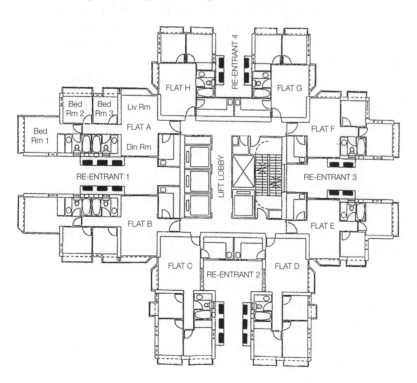

of airflow. This results in, on one hand, an elevated temperature environment at the re-entrant, and on the other hand, a greater amount of air exchange that consequently reduces the re-circulation of the kitchen exhaust.

As far as the split-type air-conditioning unit operation is concerned, an elevated temperature environment in the re-entrant significantly affects the condenser performance at the upper floor levels. The result could be an overall degradation of the capacity and efficiency of the air-conditioners of the entire building. As the condensing units may not function properly at an on-coil temperature (incoming cooling air temperature) above 45°C, the drawback is therefore not only energy wastage but also equipment malfunction. Accordingly, a HVAC engineer has to assess whether certain arrangements of the condensing units at a re-entrant will work properly even under the expected worst operating condition. This can be achieved by using CFD to study the re-entrant airflow and temperature distribution. Simplifications and assumptions have to be made to develop a steady-state computational model to facilitate the analysis[13]. Figure 2.10 shows a plot of the condenser on-coil temperature against floor level for six different condensing units in a re-entrant with 53 floor levels (hence 6×53 or 318 condensing units working together), assuming all air-conditioners operating at 50% load and in the absence of wind; outdoor temperature is 33°C. Besides, the effect of wind on the air-conditioner performance or the kitchen exhaust dilution can be predicted. Figure 2.11 shows the airflow pattern around the top part of another re-entrant in the presence of moderate wind and when the air-conditioners are turned off. Figure 2.12 shows the percentage concentration of kitchen exhaust contaminants under the steady wind situation, with Figure 2.12(a) showing the "condenser-off" situation, and Figure 2.12(b) the "condenser-on" situation. Hence with different selection of boundary conditions such as the wind conditions, the pollutant exhaust conditions, and the level of condenser operation, a detailed assessment of the air-conditioning unit performance and kitchen ventilation can be achieved[14].

3.4 Study of Airflow in Hospital Operating Theatre

The SARS crisis has led to the conversion of a positive pressure operating theatre (OT) into the negative pressure environment at a hospital of Hong Kong. This is used for the treatment of suspected or confirmed airborne infection cases, and is expected to be highly useful for special circumstances such as a return of SARS, an outbreak of Avian Influenza or other unknown epidemics.

During the airflow commissioning period of this OT, the smoke test indicated the escape of room air to the anteroom when the sliding door was operating, and this was not acceptable. Accordingly, a deflector plate was proposed to be added at one

Figure 2.10 On-coil temperature of six condensing units versus floor level at a building re-entrant

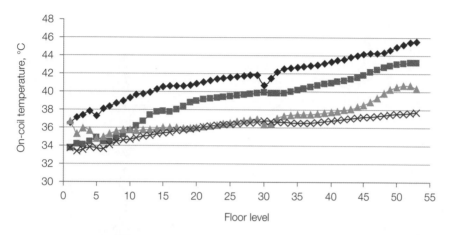

Figure 2.11 Airflow distribution at vertical plane 2 m from re-entrant end-wall under 90° moderate wind and with condenser off

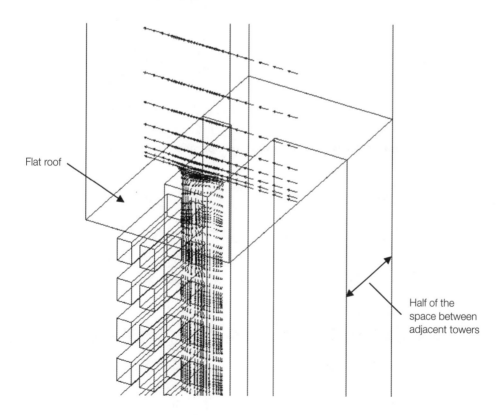

Flat roof

Half of the space between adjacent towers

Figure 2.12 Percentage concentration of kitchen exhaust contaminants under moderate wind situation

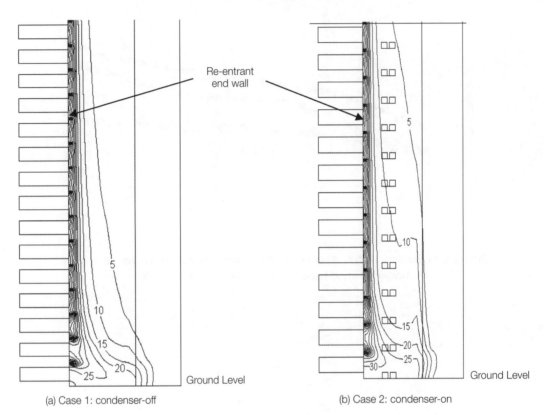

(a) Case 1: condenser-off   (b) Case 2: condenser-on

Figure 2.13 Isometric view of the operation theatre model

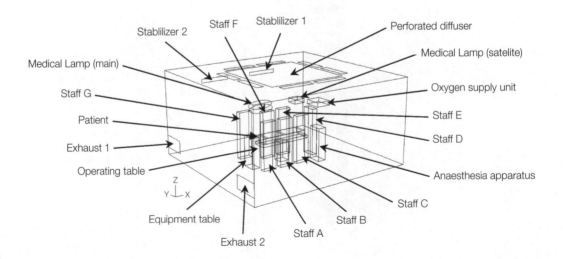

Figure 2.14 Comparison of velocity vector at vertical plane cutting across wall inlet: (a) without, and (b) with the deflector

(a) without deflector

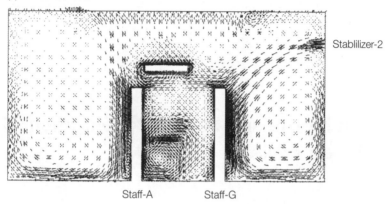

Stablilizer-2

Staff-A   Staff-G

(b) with deflector

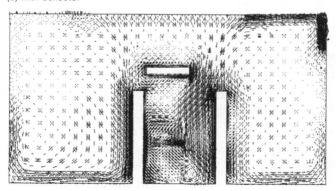

stabilizer position to prevent this air backflow. CFD was used to evaluate the effect of this deflector on the unidirectional flow at the operating table. Numerical simulations were performed with the commercial software FLUENT[15]. The standard k-ε model was adopted to simulate the flow turbulence. Only steady-state conditions were considered. Figure 2.13 shows the "Behaviour" model of the OT used in the CFD analysis. Figure 2.14 compares the velocity profiles before and after the addition of this deflector plate at Stabilizer-2. The results indicate that there is no evidence that the deflector plate will deteriorate the room airflow condition. On the contrary, the flow pattern in Figure 2.14(b) shows a better downflow at the surgical zone than in Figure 2.14(a). Hence lower bacteria concentration levels in the same zone can be observed after adding the deflector plate, from the results shown in Figure 2.15.

4 Present Needs and Future Development of Simulation Tools

4.1 Industrial Needs

The role of simulation tools in the design and engineering of buildings has been firmly established over the last two decades. Simulation is credited with speeding up the design process and enabling the comparison of a broader range of design variants, leading to better solutions. Future trends in the development of building simulation tools are driven by the needs for better support of design decisions and better quality control over performance assessments. The progression from a simple predictive tool to a more comprehensive one is welcome by most users because of the extended functionality and accuracy. For advanced simulation tools, a moderate level of proficiency requires formal training as well as hands on experience from weeks to months. The wish of the building professionals is to be able to work with some user-friendly software that can give reasonably good answers in a relatively short duration.

In principle, all software and hardware tools should only require users to have the full knowledge of the physical problem under consideration and how to use the tool. If an in-depth knowledge of the software's modelling methodology and some excessive programming works are required, then the software is not "ready" for application and improvements must be made to bring it to a working level which a practising designer who understands the physics of the phenomenon and has reasonable

Figure 2.15 Comparison of concentration of bacteria (unit: cfu/m³) released from staff at vertical plane cutting across wall inlet: (a) without, and (b) with the deflector

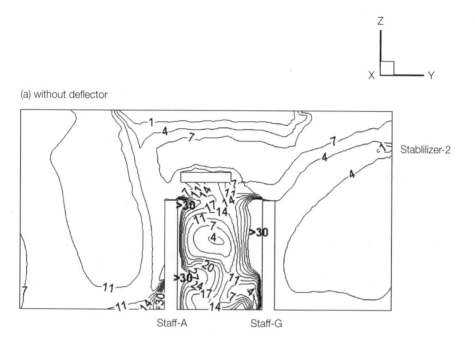

(a) without deflector

(b) with deflector

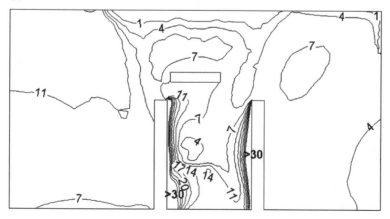

computer knowledge can use it. This may not be happening in the near future. The efficacy of dynamic thermal simulation tools in the design office or consulting practice is dependent not only on the facilities offered by the tools and the rigour of the underlying calculations, but also on the skills of the user in abstracting the essence of the problem into a model, setting up simulations and interpreting their results. Along this line, it is important for a qualified user to be familiar with the thermal properties of the building materials and plant equipment. One should understand the exact meanings of the parameters to be input to the programme, and understand the importance of time step selection. In the case of CFD tools, in spite of the strong effort for developing user-friendly packages, their efficient use still requires an expert user. It is very important to use the appropriate body geometry, boundary conditions as well as grid discretisation. How the users are able to acquire the skills mentioned above, and how simulation tools can be made more accessible are important areas of future software development.

4.2 Integrated Modelling

In the use of design tool that focuses on a single domain is likely to result in sub-optimum design solutions. It will be more appropriate, where possible, to use an integrated simulation tool throughout the design process than to use a progression of tools—from simplified to detailed—and to ignore the many theoretical discontinuities and even contradicting assumptions. While the components of a thermal model—the building, the plant, the CFD domain and the like—may be processed independently, it is better to subject them to an integrated assessment whereby the dynamic interactions are explicitly represented. A numerical study of the active/passive thermal performance of the façade-integrated PV/T collector system can help to estimate the annual availability of electricity and hot water and to predict the annual reduction of air-conditioning electricity consumption in the building. As the heat transfer at the collector glass surface relies on the approaching wind velocity, an accurate hourly performance analysis then may require an integrated model of building, plant, electric power flow and airflow.

As a matter of fact, thermal simulation and CFD programmes provide complementary information about the performance of buildings in a number of ways. By means of thermal simulation, space-averaged indoor environmental conditions, cooling/heating loads, coil loads, and energy consumption can be obtained on an hourly or sub-hourly basis for periods of time ranging from a design day to a typical year. CFD programmes, on the other hand, make detailed predictions of thermal comfort and indoor air quality, including the distribution of air velocity, temperature, relative humidity and contaminant concentrations possible. The distribution can be used to determine indices such as the predicted mean vote (PMV), the percentage of people dissatisfied (PPD) due to draft, and ventilation effectiveness. With the information from both approaches, engineers can design environmental control systems for buildings that satisfy multiple criteria.

The aim of integrated modelling is therefore to preserve the integrity of the building/plant systems by simultaneously processing all energy/mass transport paths at a level of detail commensurate with the objectives of the problem in hand and the uncertainties inherent in the describing data. The ESP-r programme for instance, is currently developing toward this direction[16]. At this end, a building and the associated systems should be regarded as being systemic, dynamic, non-linear and, above all, complex. The following is a description of an idealised comprehensive tool in our perception, as the last topic in this chapter.

4.3 Idealised Comprehensive Tool

Figure 2.16 shows the possible structure and main features of an idealised comprehensive building performance evaluation tool. The simulation programme can incorporate conversions from CAD files which are used to create a building model of arbitrary complexity. This model can be imported via the "Project Manager" (intelligent user interface), in which the form of problem/data entry can be interactive and knowledge-based. The Project Manager module gives access to support databases, simulation engine or core, which include a number of performance assessment tools and a variety of third party applications for visualisation and report generation. Its function is to co-ordinate problem definition and give/receive the data model to/from the simulation engine when the design hypothesis changes. Flexibility should be offered to allow the users more freedom to select or re-construct component models to suit various simulation tasks. The computation is processed in the simulation core, which also houses a number of inter-related sub-programmes to undertake mechanical routine calculations. The systems of equations are solved numerically, of which the process can be direct or iterative, to produce the calculation results. Selective analysis or post-processing of these results, with due considerations of human physiological responses, specified design standards, optimisation goals, etc., produces the output in suitable reporting formats.

Further extension of the software to include other building services areas like fire services or noise assessment is possible, depending on whether the integration will benefit the accuracy in the overall performance evaluation or will be to the convenience of the users. Internet may be used in the transfer of information such as simulation models, data, results, and decisions among different simulation platforms or design stations. Unavoidably, simulation tools will become more complicated and require increased technical know-how of the users to master the problem definition, simulation environment, and reporting details. The challenge of the next decade is to integrate simulation in the building design process, to be user-friendly, to increase quality control, and to exploit the explosion of communication opportunities that the Internet may offer.

Figure 2.16 Features in ideal comprehensive building performance analysis software

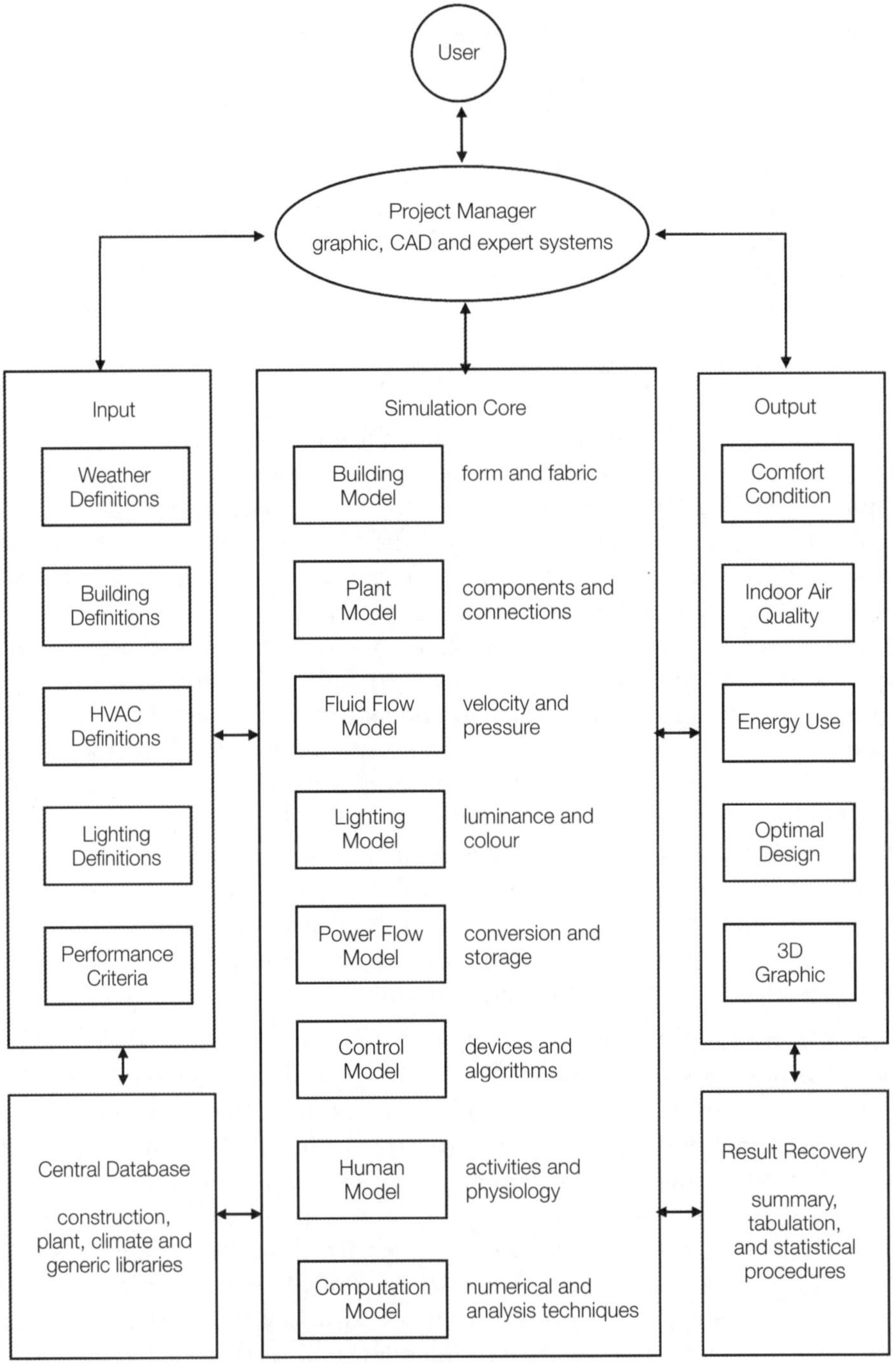

5 Conclusion

With the continuing efforts in designing and constructing high-rise intelligent and sustainable building complex in Hong Kong, the demand on the use of advanced simulation tools for helping project decisions will increase. Comparing with the laboratory testing approach, computer simulation offers the advantages of giving fast, thorough and economical outputs based on a representative virtual environment.

Different kinds of building performance evaluation tools have been introduced in this chapter, together with the application examples for illustration. While the tools are obviously becoming more and more complicated and extensive, the future trend in the software development is still governed by the basic needs of providing adequate support to the professional design/construction practices and quality control over building performance assessments. Up to now, most advanced predictive tools are regarded by the building professionals as too difficult to learn, and can only be mastered by specialists or researchers. Insufficient user-friendliness and over-demanding user-training requirements have been the major barriers in promoting the wider use of these predictive tools.

In the longer term, the advancement in computer technology in particular in computing speed and data handling will turn the building simulation tools to be profoundly comprehensive and integrative in performance evaluation. At the simulation core, the inter-related sub-programmes will support multi problem definitions, parallel data processing and online information transfer. The provision of an extensive central database and the preservation of the integrity of the building and the services systems will allow rigor mathematical treatments and hence improve the quality of the simulation results. With the improvement in the user interface, the simulation programmes will be easier to master, and hopefully will come to a stage that most users in the industry will be comfortable to work with.

Notes

1. ASHRAE. (2007). *Energy standard for building except new low-rise residential buildings*, 90.1-2007, Atlanta, US.

2. ASHRAE. (2005). *Handbook: Fundamentals*. Atlanta, US.

3. LBL. (November 1982). *DOE-2 engineers manual*. Version 2.1A, LBL-11353, Lawrence Berkeley National Laboratory, US Department of Energy.

4. LBL. (April 2008). *EnergyPlus engineering document: The reference to energyplus calculations*. Lawrence Berkeley National Laboratory, US Department of Energy, .

5. Solar Energy Laboratory. (2007). *TRNSYS 16.01.0003: A transient simulation and programme*. University of Wisconsin, Madison (WI).

6. Energy Systems Research Unit. (October 2002). *The ESP-r system for building energy simulation user guide*. Version 10 Series, ESRU Manual U02/1, University of Strathclyde.

7. Hall, I. J., Prairie, R. R, Anderson, H. E., & Boes, E. C. (1978). Generation of a typical meteorological year. *Proceedings of the 1978 Annual Meeting of the American Section of the International Solar Energy Society, 669–671.*

8. ASHRAE. (2002). *International Weather for Energy Calculations* (IWEC Weather Files) *User's Manual, Version 1.1.*

9. Crawley, D. B. (1998). Which weather data should you use for energy simulation of commercial buildings. *ASHRAE Transactions, 104*(2), 498–515.

10. Chan, A. L. S., Chow, T. T., Fong, S. K. F., & Lin, Z. (2006). Generation of a typical meteorological year for Hong Kong. *Energy Conversion and Management, 83*(1), 87–96.

11. Chow, T. T., Fong, K. F., Chan, A. L. S., & Lin, Z. (2006). Potential application of centralized solar water heating system for high-rise residential building in Hong Kong. *Applied Energy, 83*(1), 42–54.

12. Chow, T. T., Chan, A. L. S., Fong, K. F., Lo, W. C., & Song, C. L. (2005). Energy performance of a solar hybrid collector system in multi-storey apartment building. *Proceedings of IMechE Part A. Journal of Power and Energy, 219*(A1), 1–11.

13. Chow, T. T., Lin, Z., & Liu, J. P. (2002). Effect of condensing unit layout at building re-entrant on split-type air-conditioner performance. *Energy and Buildings, 34*(3), 237–244.

14. Chow, T. T., Lin, Z., Liu, J. P. (2002) Effect of condensing unit operation on kitchen exhaust at residential tower. *Architectural Science Review, 45*(1), 3–11.

15. Chow, T. T., Kwan, A., Lin, Z., & Bai, W. (2006). A computer evaluation of ventilation performance of a negative-pressure operating theater. *Anesthesia & Analgesia, 103*(4), 913–918.

16. Clarke, J. A. (2001). *Energy simulation in building design* (2nd ed). Oxford: Butterworth Heinemann.

3

IAQ, Thermal Comfort and Room Air Distribution

Mechanical ventilation, with or without heating or cooling system, is highly important for modern large buildings, especially for those tall and bulky buildings where natural ventilation is not feasible.

This chapter gives a general review on the developments of various modes of ventilation system in Hong Kong in the past decades. A brief analysis of the different ventilation design concepts for modern buildings of various types will be presented. Examples will also be given for the application of computational fluid dynamics, which could be used to study the room airflows in offices, shopping arcades, libraries and classrooms.

Zhang LIN and Wai Sun SHUM

Building Energy and Environmental Technology Research Unit
Division of Building Science and Technology
College of Science and Engineering
City University of Hong Kong

1 Ventilation

Ventilation is defined as the process of supplying and removing air by natural or mechanical means to and from any space (ASHRAE, 1992). It is vital for human health and thermal comfort (Awbi, 1991; Goodfellow, 1985; McQuiston et al., 2000), and serves multiple purposes for an indoor environment. It:

1. supplies sufficient air (oxygen) for the physiological needs of human beings/ livestock for breathing, however, this never forms a standard for ventilation since there would be acute discomfort before any danger to life arises. Hence, sufficient air movement is need for feelings of freshness and comfort;

2. supplies sufficient air (oxygen) for industrial, agricultural or other processes. For instance, a boiler plant requires air (oxygen) for combustion;

3. removes the products of respiration and bodily odour of human beings/ livestock. The outdoor air quantity required depends on the number of person and the hygiene standards;

4. removes contaminants generated by processes such as cooking, spray painting, welding, burning and wood cutting, etc.;

5. removes waste heat and water vapour etc. generated from people, lighting and equipment;

6. removes tobacco smoke and prevent the growth of molds and bacteria to achieve a good indoor air quality;

7. creates some degree of air movement which is essential for occupant's feelings of freshness and comfort; and

8. satisfies local legislative requirement.

Ventilation can be classified into two types:

1. Natural ventilation—a process of supplying and removing air by means of purpose-provided apertures (such as openable windows, doors, ventilators, shafts, etc.) and the natural forces of wind and temperature difference across the building envelope. Natural ventilation might not provide satisfactory environment for circumstances such as: internal rooms, large densely populated rooms where distribution of natural ventilation would be inadequate, buildings without sufficient openings, buildings where air extraction is required to deal with fumes and irritating gases; tall buildings where wind and stack effects make efficient natural ventilation impossible, rooms where volume per occupant

is too low for efficient natural ventilation and cases where specially close control of environment is required.

2. Mechanical ventilation—intentional, powered air exchange by a fan. It affords the greatest potential for controlling air exchange rate and air distribution within a building through proper design and operation of a ventilation system.

1.1 Determination of Ventilation Rates

1.1.1 Based on maximum allowable concentration of contaminants

The steady-state expression of decay equations is given by:

$$C_i = C_o + \frac{G}{Q} \qquad (1)$$

where C_i = maximum allowable concentration of contaminants inside occupied space

C_o = concentration of contaminants in outdoor air

G = rate of generation of contaminants inside the occupied space, l/s

Q = ventilation rate, l/s

1.1.2 Based on heat generation

The ventilation rate required to remove a given amount of heat from a building is calculated by:

$$Q = \frac{H}{c_p \rho (t_i - t_o)} \qquad (2)$$

where H = heat to be removed, W

Q = ventilation rate, l/s

c_p = specific heat of air, J/(kg K)

ρ = air density, kg/m³

t_i = indoor air temperature, K

t_o = outdoor air temperature, K

1.1.3   Based on air change rate

Various authorities publish ventilation standards, such as CIBSE Guide, ASHRAE 62-2001, government regulations, recommended ventilation rates, expressed in air changes per hour, for various situations. The ventilation rate Q expressed in m³/s is related to the air change rate by the following expression:

$$Q = \frac{V * ACH}{3600} \qquad\qquad (3)$$

where   Q   =   ventilation rate

V   =   volume of room, m³

ACH =   air change rate per hour

1.2 Mechanical Ventilation Systems

Mechanical ventilation can change the airflow through the envelope by affecting the internal pressure. This requires the installation of fans, and often ducts and auxiliary equipment to form a complete system (see Figure 3.1).

1.2.1   Supply system

In supply system, fresh air, treated or untreated, is supplied to the occupied space. There is no extraction by mechanical means. The space is slightly pressurised and excess air is vented through openings in the building envelope such as door under-cuts, window frame clearances. This type of ventilation system maintains a positive pressure within the ventilated space, which can prevent infiltration and ingress of extraneous materials. Normally this is for the assurance of oxygen supply, e.g. for industrial process, combustion process etc., in the cases of boiler room, office, and certain types of factories.

1.2.2   Extract system

The principle function of an extract system is the removal of contaminants, regardless of whether it is solid, gaseous or thermal. With such system, air is extracted from the occupied space and replaced by fresh entering from outside. The space is kept at a negative pressure, this helps in preventing the contaminants from spreading to adjoining

Figure 3.1 Mechanical ventilation system

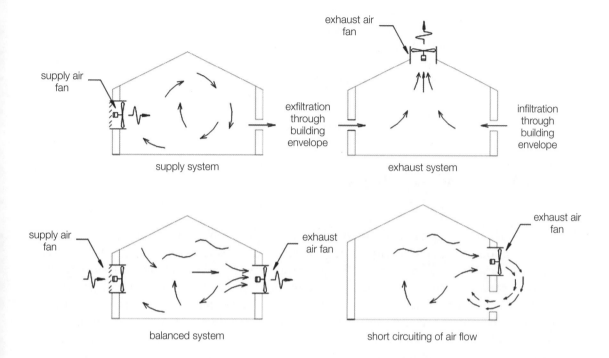

spaces. Typical examples are plant room, kitchen, toilet, refuse room, processing area, etc. The location of both the inlet and outlet openings in an extract ventilation system must not to be too close to each other to avoid short circuit. In an industrial context, local extraction of fumes or contaminants by exhaust hoods is more effective and economical.

1.2.3. Balanced system

With the combined use of supply and exhaust fans, this system is more costly. However, large openings on building envelope are not required and a close control of environment is made possible. There are two design concepts that determine the relative pressure conditions:

- a surplus of supply air over extract air so as to maintain the pressure within the building at a slightly higher level, in order to minimise natural infiltration of untreated air into the space. This concept is commonly applied in relatively clean spaces where no fumes or smells are generated, e.g. living rooms, offices, etc.

- a deficiency of supply air with respect to extract air to maintain the space at a pressure slightly lower than outside, so that any contaminant generated inside the space will therefore not spread to the adjoining spaces. Typical applications are for carparks, boiler rooms, laboratories, etc.

2 Indoor Air Quality (IAQ)

Professional health experts and building occupants generally agreed that there is a link between indoor air quality and productivity at work. Sensharma et al. (1998) assessed impacts of heating and cooling ventilation systems on thermal comfort and indoor air quality related to occupant performance and productivity. Conclusions were drawn from literature review of several authors. Dorgan (1993) reported that increases in health costs could be expected due to illness and symptoms resulting from indoor environment. Sterling and Sterling (1983) compared complaints, symptoms and performance test between a control building (mechanically ventilated) and a study building (naturally ventilated). Although complaints and symptoms occurred more often in study building than in control building, there was no significant difference in productivity between these two buildings. Berglund et al. (1990) found performance decrement in terms of number of errors related to indoor discomfort. Arora and Woods (1992) discovered that complaints, symptoms and sensitisation were factors affecting productivity, where productivity of two buildings with building related infection (BRI) or sick building syndrome (SBS) problem were defined in terms of costs of litigation, lost data from computer firm and evacuation resulting in forced absenteeism. Hall et al. (1991) asked respondents to assess the extent to which their symptoms reduced work ability or caused absenteeism. Raw et al. (1990) found a link between self-reported productivity and human responses, such as number of symptoms, comfort, temperature, humidity, satisfaction and air quality. Kroner and Start-Martin (1994) reported a significant positive relationship between affective response (overall satisfaction with the workspace for an individual worker) and occupant performance (number of files processed). In 1985, Technalysis found that more than 50% of respondents had difficulty doing work because of perceptual responses such as lack of air movement, presence of cigarette smoke, too hot in summer, too cold in winter and stagnant air. Furthermore, Sensharma et al. (1998) also stated that there were exogenous factors not related to HVAC system which had impacts on human responses in indoor environment. These factors are exogenous system parameters, exogenous exposure parameters, occupant characteristics, social characteristics and other factors. Exogenous system parameters included size of offices or workplaces. Exogenous exposure parameters included office lightning. Occupant characteristics included age, gender, education level, smoking status, allergies and job satisfaction as stated by Preller et al. (1990). Social characteristics included job title and role conflict. Other factors included availability of openable windows in rooms, office furnishings and layout, perceived crowding and work load. Since there was some evidence suggesting that adverse human responses to indoor air quality might not necessarily result in decreases of occupant productivity, exogenous factors should be taken into account to clarify the relationships between human responses to indoor air quality and occupant productivity.

Dorgan et al. (1998) summarised the effect of indoor air quality (IAQ) on employees' productivity. The first summary was on health costs and productivity benefits of improved IAQ. The second one was on medical cost reductions for specific illnesses from improved IAQ. IAQ included factors of temperature, humidity, room air motion and air contaminants. Air contaminants could be categorised as particulate or gaseous, organic or inorganic, visible or invisible, toxic or harmless, submicroscopic, microscopic or macroscopic, and stable or unstable (ASHRAE, 2005). Productivity is a measure of quality and quantity of actual accomplishments completed by employees. This productivity can be a measurable value such as sales, profits, number of errors per hour committed or actual time at work as well as a subjective value such as personal evaluation, benchmarked satisfaction of either employees or customers of supervisors' annual evaluations. Beside IAQ, other factors that affect productivity in organisations include management style, education, training, experience, salary, business stress, competition and work load. Thus, IAQ is by no means the only criterion to benchmark any productivity study. Findings showed that a one-time upgrade cost of US$87.9 billion, an annual operating cost of US$4.8 billion and a total annual benefit of US$62.7 billion resulting in a simple payback period of 1.4 years is estimated as productivity and health benefits for employees in all commercial (non-industrial) buildings studied. Contraction of diseases from indoor air of buildings is another concern. Definitions are made for sick building syndrome (SBS) and building-related illness (BRI). SBS is symptoms contracted while staying in indoor environments. These symptoms include eye, nose and throat irritation, dryness of mucous membranes and skin, nose bleeds, skin rash, mental fatigue, headache, cough, hoarseness, wheezing, nausea and dizziness (ASHRAE, 2005). BRI is defined as specific recognisable disease that can be clearly related to chemical, infectious or allergic agents in buildings. Hypersensitivity pneumonitis, humidifier fever, occupational asthma and Legionella infection are often typical BRI. Indicators of BRI include cough, chest tightness, fever, chills, sinus congestion, headaches and muscle aches. These symptoms usually require prolonged recovery times after leaving a building.

Basically there are three major types of indoor air pollutants: airborne pathogens, volatile organic compounds and carbon dioxide concentration level.

2.1 Airborne Pathogens

Li et al. (2007) demonstrated a definitive association between transmission of airborne infection viruses and ventilation of buildings. These infection viruses include but not limit to severe acute respiratory syndrome (SARS) epidemic, avian influenza (H5N1) pandemic, measles, tuberculosis, chickenpox, influenza and smallpox. Although there are strong and sufficient evidence that link the relationships among ventilation, air movement in buildings and transmission or spread of infectious diseases, no specific ventilation requirements are imposed for various occupancy types of buildings which include hospitals, schools, offices, homes and isolated rooms regarding to spread of airborne infectious diseases.

Amoy Gardens, a high-density middle-class housing estate in Hong Kong, was the most seriously affected location during the 2003 Severe Acute Respiratory Syndrome (SARS) outbreak. From end of March 2003 to middle of April 2003, a total of 321 people in the estate were reported with SARS infection. All residents were subsequently relocated to Lei Yue Mun and MacLehose Holiday Camps for quarantine. Residents were only allowed to return after decontamination of the estate. As a result of the outbreak in Hong Kong as of 13 May 2003, more than 23 people lost their lives and more than 800 were sickened. Subsequently, Yu et al. (2004) made use of epidemiologic analysis, experimental studies and computational fluid dynamic simulations, which supported the probability of SARS being spread through airborne transmission at Amoy Gardens after visited by an infected patient. No mechanical ventilation was in operation due to moderate weather although exhaust fans were turned on at the time of suspected infection. Li et al. (2004) applied multi-zone modelling (a zone was a flat or a room) together with computational fluid dynamics to simulate the transmission of the SARS virus among flats at Amoy Gardens. Results suggested the possible airborne transmission of the virus which led to the need of improving indoor air quality and ventilation design in buildings.

2.2 Volatile Organic Compounds

VOCs particularly include toluene, benzene and formaldehyde. Formaldehyde usually emanated from chairs, tables and wood paneling. Benzene is a major component of paints and resins which has been classified by international agency for cancer research to be carcinogenic to humans. Toluene can be found in paints, paint thinners and adhesives.

Sundell (1996), Hetes et al. (1995) and Wolkoff et al. (1992) have confirmed the positive association between photocopiers and sick building syndrome. Pollutants emitted from photocopiers include volatile organic compounds (VOCs), ozone, respirable particles, formaldehyde and nitrogen dioxide. Survey conducted by Brown (1999) shows that there is a significant increase in VOCs emission by the photocopier in operating mode when comparing with idle mode. Nevertheless, emissions of ozone, nitrogen dioxide and formaldehyde are small.

Personal computers (PCs) penetration to home and office environment have been revolutionary in recent decades. In a study performed by Bakó-Biró et al. (2004), percentage of people dissatisfied (PPD) and sensory pollution load in an office are both increased when PCs are present compared to the condition without PCs, although no SBS symptoms are reported. A range of VOCs is detected during the presence of PCs including phenol, toluene and formaldehyde.

2.3 Carbon Dioxide Concentration Level

There is a growing interest in the impact of indoor air quality on work performance in commercial buildings. In addition, some studies divulged that ventilation rate above 10 l/s per person reduces SBS symptoms. Results obtained by Federspiel et al. (2004) on workers' performance in a call centre under various carbon dioxide concentration differences between indoor and outdoor air revealed that work performance is statistically the same at low and high ventilation rates, where ventilation rates are in inverse proportion to differential indoor and outdoor carbon dioxide concentration levels. However, work performance at intermediate ventilation rates is worse by about 2%.

Shendell et al. (2004) conducted a study on the relationship between indoor CO_2 concentration and student absenteeism in Washington and Idaho. Generally speaking, lower ventilation rate corresponds to higher indoor CO_2 concentration. ASHRAE (2005) specifies a minimum ventilation rate of 7.5 l/s per occupant for classrooms. Indoor CO_2 concentration exceeds outdoor concentration because of metabolic production of CO_2 by building occupants. For a ventilation rate of 7.5 l/s per occupant and a typical outdoor CO_2 concentration of 350–400 ppm, indoor CO_2 concentration of 1,000 ppm has been used as an informal dividing line between "adequate" and "inadequate" ventilation according to ASHRAE (2005). Statistical analysis performed by Shendell et al. (2004) is significant which indicates that a 1,000 ppm indoor CO_2 concentration above outdoor

CO_2 concentration has led to 10–20% relative increase in student absence. They also reason that this is probably attributed to an increase in airborne infectious particles under higher indoor CO_2 concentration.

School children's performance link to insufficient indoor air ventilation is investigated by Shaughnessy et al. (2006). Carbon dioxide concentration in classroom is monitored for students' performance on math and reading aptitude tests. Ventilation rates are categorised into four groups, (i) less than 2.25, (ii) between 2.25 and 3.5, (iii) between 3.5 and 4.5, (iv) greater than 4.5 l/s per person. Statistical results indicate that reading and math scores are both in direct proportion to ventilation rates. That is, the higher the ventilation rate, the higher the test scores.

Erdmann and Apte (2004) investigate the correlation between indoor carbon dioxide concentration and building related symptoms (BRS) or sometimes called sick building syndrome (SBS). Indoor carbon dioxide concentration can be considered as a surrogate for other indoor pollutants, but it is not expected to cause health symptoms directly. ASHRAE (2005) has recommended a minimum office ventilation rate of 10 l/s/person which corresponds to 870 ppm CO_2 concentration indoor when outdoor CO_2 concentration is 350 ppm and indoor CO_2 generation rate of 0.31 l/min/person. Statistical results show that differential of indoor and outdoor CO_2 concentration is not significantly related to BRS. Nevertheless, elevated indoor CO_2 concentration increases prevalence of mucous membrane and lowers respiratory building related symptoms such as dry eyes, sore throat, nose sinus, sneeze and wheeze.

Based on mass balance calculations by Apte et al. (2000), the lowest minimum ventilation rate acceptable corresponds to a maximum steady state indoor CO_2 concentration of 1,000 ppm when outdoor CO_2 concentration is 350 ppm and a CO_2 generation rate per person is 0.31 l/min. They also mention about a survey carried out by Seppänen et al. (1999) which shows that about one-half of the studies are statistically significant and indicates a positive correlation between CO_2 concentration and sick building syndrome (SBS), where SBS includes upper respiratory and mucosal symptoms, dry, itchy, sore, irritated eyes, noses and throats, wheezing, headache, drowsiness, tiredness, mental fatigue, dizziness and skin dryness or itching. Several other studies have found that increasing ventilation rates to 20 Ls–1 significantly reduce SBS.

Mi et al. (2006) discovers that NO_2 and CO_2 concentrations increase the occurrence of asthma. Also, it is generally agreed that ozone reacted with other indoor pollutants to form more irritative compounds. In addition, they find that indoor formaldehyde is not significantly connected to respiratory symptoms. Sundell (1994) remarks that the use of mechanical ventilation to dilute indoor air pollutants by outside air is a practical method to provide acceptable indoor air quality. In a school with low ventilation rate compared with another one with higher ventilation rate studied by Wålinder et al. (1997), airborne pollutants such as carbon dioxide, respirable dust, total volatile organic compounds, moulds and bacteria are all measured significantly higher. Results also suggested that exposure to high level of indoor air pollutants may affect airways and causes swelling of nasal mucosa.

3 Thermal Comfort

Thermal comfort variables in an occupied zone include metabolic rate, clothing, air velocity, air temperature, air temperature stratification, radiant temperature, radiant temperature asymmetry, relative humidity and turbulence intensity. A mathematical model is established by Fanger et al. (1989) which predicts the percentage of people dissatisfied due to draft as a function of mean air velocity, turbulence intensity and air temperature.

3.1 Air Temperature

Chan et al. (1998) studied a large-scale survey of thermal comfort conditions in Hong Kong office premises which is sparked by the sub-tropical climate and the dense population in Hong Kong. According to ASHRAE Standard 55, typical indoor air temperature is set to 24°C. During energy crisis, Hong Kong government has encouraged a temperature of 25.5°C in all government office buildings. When crisis were over, indoor air temperature is market driven and has been set to 23.5°C for Grade A buildings, 24°C for Grade B buildings and 25.5°C for Grade C buildings, where Grade A buildings have the highest quality with Grade C buildings the lowest. Thirteen office premises are under survey in the sample. Questionnaires were distributed to office employees with responses taken in summer and winter times. In the conclusion of the survey, the preferred office temperature in summer is found to be 22.5°C which is one degree Celsius lower than the neutral temperature 23.5°C, where neutral temperature is one of the seven categories of the seven-point scale in thermal sensation assessment ranging from cold, cool, slightly cool, neutral, slightly warm, warm to hot. Lower office temperature is preferred in Hong Kong compared to other regions around the world, which was probably due to additional thermal insulation of clothing worn. This survey outcome provides a better insight for thermal comfort requirement to be imposed on HVAC systems design criteria.

3.2 Air Humidity

Fang et al. (2000) studied about the effect of temperature and humidity on the perception of indoor air quality. Three experimental setups were scrutinised, facial exposure experiment which was carried out in a test box made of glass ($1.005 \times 0.25 \times 0.22$ m³), whole body exposure experiment which was carried out in a climate chamber made of stainless steel ($3.6 \times 2.5 \times 2.55$ m³) and field study which was carried out in an office room (36 m²) equipped with air-conditioning system. All three experiments give conclusion that although temperature and humidity do not affect odour perception of indoor air, they do significantly influence acceptability of indoor air. Furthermore, acceptability of air is decreasing linearly with increasing enthalpy of air. Results shown in this chapter emphasised that temperature and humidity have tremendous impact on perceived air quality, but some ventilation standards, such as CEN 1998 and ASHRAE 2001, do not include the impact of air temperature and humidity for indoor air quality.

Two models for predicting discomfort due to skin humidity and insufficient respiratory cooling are proposed by Toftum and Fanger (1999). The skin humidity model predicts discomfort as a function of relative humidity of skin. The respiratory model predicts discomfort as a function of driving forces for heat loss from temperature and humidity of surrounding air. At present, air humidity affecting comfort is not completely known. However, air humidity does affect evaporation of moisture from human skin and respiratory tract. Thus, it has an influence on human heat loss and thermal sensation. The predominant factors for humidity discomfort are basically because of high level of skin humidity and insufficient cooling of mucous membranes in upper respiratory tract. Skin humidity is determined by human sweat and skin wetness when in contact with clothing. Insufficient respiratory cooling is caused by high air temperature or high air humidity when air is inhaled through mucous membranes in upper respiratory tracts. High air humidity causes condensation in ducts which results in growth of fungi or other microorganisms that would trigger respiratory infections, eye irritations and skin rashes. Studies show that upper humidity limit based on skin humidity is less restrictive but the limit based on respiratory discomfort is far more restrictive.

Experiments on thermal comfort at high humidity are performed by Fountain et al. (1999) in a climate chamber. Studied subjects were exposed to an environment ranging from 20°C (68°F) / 60% relative humidity (RH) to 26°C (78.8°F) / 90% RH. In the documents of the American Society of Heating and Ventilating Engineers (ASHVE), both design and operation humidity standards for HVAC systems in the U.S. before 1915 almost exclusively focused on ventilation rates without mentioning humidity for different classes of buildings. In 1915, the society introduced the Code of Minimum Requirements for ventilation with an upper relative humidity limit of 50% being desirable but not mandatory. Since then, both ASHVE and ASHRAE issued maximum humidity range for HVAC systems between 50% and 75%. In a recent paper,

Berglund (1995) analyses the JB Pierce foundation study of humidity effects on thermal comfort carried out by Berglund (1989) and concludes that the judgment of whether an environment is thermally acceptable or not depends on both thermal sensation and perceived skin moisture. This implies that at a constant temperature, increasing humidity increases discomfort. Tanabe et al. (1994) suggest that at high humidity levels (80% RH), the perception of thermal sensation is not a reliable indicator of thermal comfort and concludes that thermal sensation is inappropriate to predict thermal comfort at high levels of humidity. Based on these studies, humidity and thermal comfort do not show a clear relationship. However, other factors such as temperature and clothing are proven to be more reliable indicators for thermal comfort. The mechanism that humidity affects comfort is not known. Thermal discomfort due to humidity might result from a clinging sensation of clothing on wet skin. The study concludes that although the experimental study does not indicate that there are no effects of humidity on thermal comfort, the study is rather unable to clearly find that if there are effects. In particular, it is found that 90% RH condition is least favourable, 80% RH condition is not apparently worse than 60% or 70% RH condition, and 70% RH condition is rated more favourable than 60% RH condition.

4 Room Air Distribution

Different ventilation modes provided by Heating, Ventilation and Air Conditioning (HVAC) systems affect room air distribution and hence the efficiency to dilute indoor air with fresher outdoor air. Ventilation refers to the process of introducing and distributing outdoor air into a building or a room. The amount of air circulated per unit time is termed ventilation rate. Air distribution pattern in a ventilation process is a mode of air movement within buildings. Ventilation process can involve airflow by either natural thermal buoyancy or convective air stream generated by fans. Meanwhile, air movement can be controlled and affected by physical obstacles in buildings or movements of occupants. Zhivov and Rymkevich (1998) analyse HVAC systems adopting mixing ventilation (MV) and displacement ventilation (DV) strategies for heating and cooling in more than 200 restaurants across different US climates. Analysis shows that displacement ventilation allows better IAQ. The increase of air supply in Miami and Albuquerque restaurants has resulted in the increase of heating and cooling energy consumptions. But the increase of air supply for restaurants in other areas has resulted in the decrease of cooling energy consumption. HVAC system operated in Phoenix, Minneapolis and Seattle restaurants with variable air supply can cut cooling energy consumption by as much as 50%. Displacement ventilation systems reduce cooling energy consumption, although heating energy consumptions are augmented.

Evaluation of ventilation effectiveness and thermal comfort of various industrial ventilation schemes are carried out by scale model experiments. The objective is to compare the performances of supply and return diffusers located at different heights for a range of flow rates and heat loads. Small scale model experiments are tested to predict full scale conditions by keeping the same Reynolds number, the Archimedes number, the Prandtl number and the Schmidt number. The largest ventilation effectiveness occurred for a low supply/high return configuration, followed by high supply/high return configuration and then low supply/low return configuration. Increasing the number of diffusers in occupied zone provide increasing ventilation effectiveness. For a given configuration, ventilation efficiency is generally increasing when heat load is increasing and/or flow rate is decreasing. Thermal comfort depends on diffuser configurations as well as activity level of workers. Most configurations produce acceptable thermal comfort for seated workers, but unacceptable thermal comfort for workers with increased activities and clothing level. Ventilation modes under investigation are mixing ventilation, in which fresh cool air mixed with existing room air, and displacement ventilation, in which fresh cool air displaced existing room air.

There are basically four ventilation modes for air-conditioned non-industrial indoor space. They are mixing ventilation, displacement ventilation, task/personalised ventilation and mostly recently stratum ventilation.

4.1 Mixing Ventilation

Mixing ventilation is a traditional mode of ventilation and is still widely used today. It can take care of both heating and cooling with highly varying load patterns. Mixing is achieved by supplying ventilation air as a high velocity jet to entrain air already in the room. Mixing ventilation can provide comparably uniform air temperature distribution in occupied zone. However, it also leads to problems such as poor IAQ, air draft in occupied zone and some other thermal discomfort. In addition, it normally consumes more energy than displacement ventilation.

4.2 Displacement Ventilation

Displacement ventilation has been used quite commonly in Scandinavia during the past 25 years as a mean of ventilation in industrial facilities to provide better IAQ and to save energy. More recently, its use has been extended to ventilation of offices, classrooms, commercial buildings and other non-industrial premises. In contrast to mixing ventilation, buoyancy forces (induced by heat sources) govern the airflow in displacement ventilation. Because airflow is thermally driven, this mode of ventilation is only satisfactory when excess heat is needed to be removed. In a room ventilated by displacement ventilation mode, air quality in breathing zone is usually better than that in mixing ventilation operated with the same airflow rate. Ventilation efficiency of a displacement ventilated room is also significantly better than that of mixing ventilated room (Awbi HB, 1991). It is found that an increase in ventilation efficiency is often achieved at the breathing level (Lin et al., 2005a). But there are more severe restrictions due to thermal discomfort in a displacement system than in a mixing system. ISO 7730 (1984) recommends a maximum air temperature gradient of 3 K between 1.1 m and 0.1 m above floor which corresponds to a percentage people dissatisfied of 5%. However, the ASHRAE (1992) Standard recommends the same temperature gradient between 1.7 m and 0.1 m above floor when related to a standing person, or it will cause thermal discomfort otherwise. So supply air temperature and airflow rate must be carefully studied to assure proper temperature distribution and velocity distribution. Often the lower part of a space embraces few heat sources. Inevitably, a displacement ventilation system cools the lower part excessively, resulting in discomfort and energy waste.

Yuan et al. (1999) suggest that at design stage, performance of displacement ventilation is evaluated by thermal comfort level, indoor air quality (IAQ), energy consumption, initial cost and maintenance costs of HVAC systems. Criteria to evaluate thermal comfort further broke down into air temperature distribution, percentage of people dissatisfied due to draft (PD) and predicted percentage of people dissatisfied for thermal comfort (PPD). Contaminant concentration distributions and mean age of air are often good indicators for IAQ. Air temperature distribution and ventilation rate affect energy consumption. All these performance parameters are determined by thermal and flow boundary conditions such as size and geometry of space, heat sources and contaminant sources. With displacement ventilation, airflow velocity is lower than 0.2 m/s. Temperature difference between head and ankle level of a sedentary occupant is less than 2 K. Both PD and PPD are less than 15%. Compared to mixing ventilation, displacement ventilation provides better IAQ if contaminant sources are associated with heat sources. Other contaminants may come from volatile organic compounds evaporated from building materials, which might not provide better IAQ with displacement ventilation. In addition, mean age of air is younger and ventilation effectiveness is higher.

4.3 Task/Personalised Ventilation

As far as IAQ of breathing zone and energy efficiency are concerned, task/personalised ventilation is the most effective. It may be used to remove excessive cooling load and maintain a comfortable indoor environment. Task/personalised ventilation systems supply air through nozzles located near occupants (e.g. at an edge of a desk). Potential draft exists because of short distance between supply gears and occupants. A field study finds that workers in a task ventilated office feel satisfied with thermal conditions because they can individually control local environment (Bauman et al., 1993). Occupants can control temperature, flow rate and direction of air from nozzles. Measurements conducted by Faulkner et al. (1993, 1995) show that age of air at breathing level with task/personalised ventilation is approximately 30% younger than that with mixing ventilation. However, application of task/personalised ventilation depends much on indoor furnishings. On one hand, it is difficult or expensive to equip nozzles and connect duct in various indoor spaces, especially to keep up with the paces of re-partitioning. On the other hand, some occupants do not usually stay in a fixed location, for instance, customers in a retail shop. These limitations restrict the use of task/personalised ventilation.

Raised flooring systems have drawn much attention recently. Task/personalised ventilation system can involve underfloor air-conditioning. Power and data cables housed under floor cavities can be easily accessed and modified to accommodate changes in the use of space above floor. These cavities can act as supply air plenums for conditioned air to get to occupants through the floor cavities. Meanwhile, return air plenums continue to be housed in ceiling cavities. A study has been discussed by Loudermilk (1999) about the improvement of space ventilation, reduction of installing costs and operating costs for underfloor air-conditioning systems. In underfloor ventilation systems, indoor temperature and contaminant levels are increasing from floor to ceiling. Comparing to mixing ventilation, indoor temperature and contaminant levels stay practically the same from floor to ceiling. Generally speaking, an underfloor air-conditioning system consists of two zones, mixing and displacement zones. Mixing zone is the well mixed layer of air measured about 1.2 metres from the floor. From the top of this layer to ceiling, it is the displacement zone because air movement resembles displacement ventilation. The advantages of using underfloor ventilation include improved space ventilation with which occupants can adjust and control over the conditioned air for personal use, reducing mechanical equipment costs by which reduction of ductwork beyond terminal units can be achieved, enhanced space flexibility with which changes in space by relocating furniture and partitions becomes easier.

Loomans (1999) draws conclusions of experimental and numerical studies on desk displacement ventilation (DDV) concept, which is a combination of displacement ventilation with task conditioning. In task conditioning system, air supply diffuser that draws cool air from floor plenum is situated near an occupant. In the DDV system, air supply diffuser that draws cool air from floor plenum is situated below desktops against the back of desks towards an occupant. With these configurations, DDV inherits the features of higher ventilation effectiveness compared to mixing ventilation and more control of thermal comfort like those of task conditioning. However, the studies imply that DDV is not feasible unless special attention is given to room configuration.

4.4 Stratum Ventilation

Specific limitations exist in mixing ventilation, displacement ventilation and/or task/personalised ventilation. To overcome these problems, a new ventilation mode has been developed. The new system should be able to provide good IAQ in breathing zone, to have minimum temperature difference between head and ankle levels to obtain thermal comfort in occupied zone, to equip duct and diffusers conveniently and to have high energy efficiency.

The underlying principle of displacement ventilation implies that in an air-conditioned room, the conditions of IAQ and thermal comfort beyond occupied zone and beneath breathing zone (approximately $z > 2$ m and $z < 0.8$ m, where z is measured height from the floor.) are of little interest. For conventional ventilation modes, breathing air is transported by boundary layer around the body of an occupant and air quality is a weighted average air quality in a room from the breathing level to floor level. Ventilation efficiency would be maximised if air is supplied directly into breathing zone and air form a well controlled "fresher air layer" to fill a breathing zone. The thickness of fresher air layer depends on the nature of occupancy. Meanwhile, a quasi-stagnant zone is also formed between breathing zone and floor (approximately $0 < z < 0.8$ m). If the temperature within quasi-stagnant zone is not as low as displacement ventilation, the problem of "cold ankles" can be solved. Energy is also saved by avoiding over-cooling of the lower part of a room.

Stratum ventilation works by creating a layer of fresher air in occupants' breathing zone. This is done by placing supply inlets at the side-wall of a room slightly above the height of occupants, standing or sitting, depending on application. Fresh air enters the room and gradually loses momentum farther away from supply. Fresh air supply is sufficient to provide adequate air of young age. The range of face velocity and locations of air supply and exhaust should be carefully optimised to break boundary layer around the body of an occupant to minimise risks of draft and cross contamination. A knowledge based system which consists of a database related to the drop distance of an air jet based on the Archimedes number (Ar) should be developed to decide the appropriate air supply velocity (Leonard and McQuitty, 1986; Zhang and Strom, 1999). The thickness of fresh air layer required depends on the nature of occupancy. At the same time, a quasi-stagnant zone is also formed between breathing zone and floor (say $0 < z < 0.8$ m). Temperature within quasi-stagnant zone should be reasonably controlled. In addition, the supply of air at this level increases the convection effect of heat and helps to displace contaminants into unoccupied zone. It therefore brings fresher air to breathing zone than that of conventional modes of ventilation. Indoor air should be mixed well and air temperature gradient should be low in order not to cause thermal discomfort. Because air is supplied directly to the breathing zone, less fresh air bypasses occupants. Thus, there are also possibilities to reduce fresh airflow rate for energy saving.

Since air is supplied directly to the breathing zone, air supply temperature should be usually above 20°C. This implies that condensing temperature of refrigerating plant can also be elevated accordingly, which results in higher coefficient of performance (COP).

5 Numerical Modelling

Although CFD techniques have been widely used for HVAC computation, Chen (1997) has showed some uncertainties on using turbulence models. Direct numerical simulation (DNS) applying to Navier-Stokes equations is not preferred due to the incapacity of present computing powers to resolve fine structures of turbulence using very fine grids. Since most kinds of flow are turbulent and the major interest is on mean average parameters of a flow, hence the less accurate but more efficient turbulence models are adopted. However, a turbulence model might perform well in one flow but poorly in another. Thus, experimental validation is always required to ensure the suitability of the turbulence model for each given flow.

5.1 Turbulence Models for Room Air Distribution

Direct numerical simulation (DNS) of laminar indoor airflow using Navier-Stokes equations is practically implausible because it requires an extremely high number of grid cells in capturing turbulence fluctuations and dissipation of turbulence fluctuations. As a result, simulation using turbulence models comes into play. Different turbulence models are discussed by Nielsen (1998) when applied to different airflow situations. Zero-equation model can be used for prediction of low level turbulence flow. Two equations k-ε model is used to predict room airflow under displacement ventilation. Low Reynolds number (LRN) model is used to predict evaporation-controlled emissions from building material at or near walls. Large eddy simulation (LES) is used to average small-scale eddies to save computing time. Indoor airflow regimes are usually described by turbulence. Suppose room airflow velocity $\hat{u}$ is given by

$$\hat{u}(x, y, z, t) = u(x, y, z) + u'(x, y, z, t),$$

where u is the mean air velocity and u' is perturbed velocity from the mean velocity. If the above equation is average over a time period from t_1 to t_2, then

$$u(x, y, z) = \frac{1}{t_2 - t_1} \int_{t_1}^{t_2} \hat{u}(x, y, z, t) dt \ .$$

Obviously, $\dfrac{1}{t_2 - t_1} \displaystyle\int_{t_1}^{t_2} u'(x, y, z, t) dt = 0$. In addition to the physical viscosity μ of airflow, there exists the turbulent viscosity μt due to turbulence. Hence, the effective viscosity μ_{eff} of average flow equations becomes

$$\mu_{eff} = \mu_t + \mu.$$

For zero-equation turbulence model,

$$\mu_t = \text{constant} \times \mu,$$

or

$$\mu_t = \text{constant} \times \rho U H,$$

where ρ is density, U is characteristic velocity and H is characteristic length (room height). For two equations k-ε turbulence model, a transport equation for turbulent kinetic energy k and a transport equation for dissipation of turbulent kinetic energy ε are involved. According to Launder and Spalding (1974), k-ε turbulent model is written as

$$\frac{D(\rho k)}{Dt} = \frac{\partial}{\partial x_i} \left[\left(\mu + \frac{\mu_t}{\sigma_k} \right) \frac{\partial k}{\partial x_i} \right] + P + G - \rho\varepsilon - 2C_5\mu \left(\frac{\partial\sqrt{k}}{\partial x_i} \right)^2 ,$$

$$\frac{D(\rho\varepsilon)}{Dt} = \frac{\partial}{\partial x_i} \left[\left(\mu + \frac{\mu_t}{\sigma_\varepsilon} \right) \frac{\partial\varepsilon}{\partial x_i} \right] + C_1 f_1 \frac{\varepsilon}{k}(P + C_3 G) - C_2 f_2 \rho \frac{\varepsilon^2}{k} + 2C_4 \frac{\mu\mu_t}{\rho} \left(\frac{\partial^2 u_i}{\partial x_i \partial x_j} \right)^2 ,$$

where

$$P = \mu_t \frac{\partial u_i}{\partial x_i} \left[\left(\frac{\partial u_i}{\partial x_i} + \frac{\partial u_j}{\partial x_i} \right) \right],$$

and the buoyancy term G can be obtained by

$$G = \beta g_i \frac{\mu_t}{\sigma_t} \frac{\partial T}{\partial x_i} .$$

Turbulent viscosity μ_t is obtained from

$$\mu_t = \rho \, C_\mu f_\mu \frac{k^2}{\varepsilon} .$$

Constant parameters involved are $C_1 = 1.44$, $C_2 = 1.92$, $C_3 = 0.09$, $\sigma_k = 1.0$ and $\sigma_\varepsilon = 1.3$. For conventional k-ε turbulence model,

$$f_1 = f_2 = f_\mu = C_3 = 1 \text{ and } C_4 = C_5 = 0.$$

When airflow is at or near solid boundary of walls, low Reynolds number k-ε turbulence model is then required. According to Launder and Sharma (1975),

$$f_1 = C_3 = C_4 = C_5 = 1.0,$$

$$f_2 = 1.0 - 0.3\exp(-R_t^2),$$

$$f_\mu = \exp[-3.4/(1 + R_t/50)^2],$$

where the turbulent Reynolds number is defined as

$$R_t = \frac{\rho k^2}{\mu \varepsilon}.$$

Another type of k-ε turbulence model can be formulated if

$$f_\mu = f_{R_t} f_b,$$

where

$$f_{R_t} = \exp[-3.4/(1 + R_t/50)^2],$$

$$f_b = \begin{cases} 0.0 & \text{for } b \leq -10 \\ 1 + b/10 & \text{for } -10 < b < 0, \\ 1.0 & \text{for } b \geq 0 \end{cases}$$

$$b = G/\varepsilon.$$

Alternatively, suppose room airflow velocity $\hat{u}$ is given by

$$\hat{u}(x, y, z, t) = u(t) + u''(x, y, z, t),$$

where u is the mean air velocity and u'' is perturbed velocity from the mean velocity. If instead of filtering over a time period, filtering is done over a local differentially small spatial volume $\Delta x \Delta y \Delta z$, then

$$u(x, y, z, t) = \frac{1}{\Delta x \Delta y \Delta z} \iiint \hat{u}(x, y, z, t) dx dy dz.$$

Obviously, $\dfrac{1}{\Delta x \Delta y \Delta z} \iiint u''(x, y, z, t) dx dy dz = 0$ in a local volume $\Delta x \Delta y \Delta z$. This kind of spatial filtering applied to Navier-Stokes equations is the basis of large eddy simulation, see Murakami (1988). To close the system of governing equations, expression for subgrid-scale Reynolds Stresses (SGS) can be applied.

Emmerich and McGrattan (1998) performs large eddy simulation (LES) with Smagorinsky subgrid-scale (SGS) model to solve a three-dimensional room ventilation airflow problem. Computed results are in good agreement with measured data. Kurabuchi et al. (1990) develops a CFD programmme called EXACT3 that employs a k-ε two equations turbulence model to predict indoor airflow. Recently, McGrattan et al. (1994) develops a CFD programmme (NIST-LES3D) at the National Institute of Standards and Technology (NIST) that embraces LES to simulate smoke transport during a fire in an enclosure. Chen and Chao (1996) find that Reynolds stress model is best for turbulent buoyancy plume, while renormalisation group (RNG) k-ε model is best suited for displacement ventilation. Sakamoto and Matsuo (1980) study isothermal flow in a room using both k-ε model and large eddy simulation model. The overall results are in good agreements with experimental data. Although the models do not agree well with experimental results near exhaust outlet, LES corresponded better with experimental results around supply inlet. Davidson and Nielsen (1996) and Bennetsen et al. (1996) report LES with dynamic subgrid model are better performed than LES with Smagorinsky subgrid model when compared with measured data.

In fact, McGrattan et al. (1999) present a LES simulation of fire smoke movement in buildings by applying an approximated form of Navier-Stokes equations. Fire-related phenomena such as radiative heat transfer, flame spread and sprinkler spray dynamics have been added to the model to enable large-scale fire numerical experiments. The mathematical model is derived directly from the Navier-Stokes equations and the ideal gas law. From conservation of mass,

$$\frac{\partial \rho}{\partial t} + \nabla \cdot \rho \mathbf{u} = 0. \tag{1}$$

From momentum transport,

$$\rho \left(\frac{\partial \mathbf{u}}{\partial t} + \frac{1}{2} \nabla \cdot |\mathbf{u}|^2 - \mathbf{u} \times \boldsymbol{\omega} \right) + \nabla p - \rho \mathbf{g} = \nabla \cdot \boldsymbol{\sigma}. \tag{2}$$

From energy transport,

$$\rho c_p \left(\frac{\partial T}{\partial t} + \mathbf{u} \cdot \nabla T \right) - \frac{dp_0}{dt} = \dot{q} + \nabla \cdot k \nabla T. \tag{3}$$

From ideal gas law,

$$p_0(t) = \rho R T. \tag{4}$$

The symbols have their usual fluid dynamical meaning, ρ is density, u is velocity vector, ω is vorticity, p is pressure, g is gravity vector, c_p is constant pressure specific heat, T is temperature, k is thermal conductivity, t is time, $\dot{q}$ is prescribed volumetric heat release, R is gas constant with $R = c_p - c_v$, c_v is constant volume specific heat and σ is standard stress tensor for compressible fluids. By taking total derivative of Equation (4) and substituting expressions from Equations. (1) and (3),

$$p_0 \nabla \cdot \mathbf{u} + \frac{1}{\gamma} \frac{dp_0}{dt} = \frac{\gamma - 1}{\gamma} (\dot{q} + \nabla \cdot k \nabla T) \tag{5}$$

where $\gamma = c_p / c_v$. Integrating Equation (5) over the entire domain Ω yielded

$$p_0 \int_{\partial\Omega} \mathbf{u} \cdot \mathbf{dS} + \frac{V}{\gamma} \frac{dp_0}{dt} = \frac{\gamma - 1}{\gamma} \left(\int_{\Omega} \dot{q} dV + \int_{\partial\Omega} k \nabla T \cdot \mathbf{dS} \right), \tag{6}$$

where Ω is the interior of computational domain, $\partial\Omega$ is the boundary of computational domain, V is the volume of the computational domain, dS = ndS is differential surface area vector of the enclosed volume of the computational domain and n is unit outward normal on dS. Introducing the background pressure p0(t) and background density $\rho_0(t)$

$$p_0(t) = \rho_0 R T_0. \tag{7}$$

Furthermore, temperature $T(t)$ and density $\rho(t)$ can be expressed as

$$T = T_0(t)(1 + \tilde{T}), \tag{8a}$$
$$\rho = \rho_0(t)(1 + \tilde{\rho}), \tag{8b}$$

where the perturbation values of $\tilde{T}$ and $\tilde{\rho}$ are related by

$$(1 + \tilde{T})(1 + \tilde{\rho}) = 1. \tag{9}$$

From adiabatic process,

$$\frac{\rho_0}{\rho_\infty} = \left(\frac{p_0}{p_\infty} \right)^{\frac{1}{\gamma}}, \tag{10a}$$

$$\frac{T_0}{T_\infty} = \left(\frac{p_0}{p_\infty} \right)^{\frac{\gamma - 1}{\gamma}}, \tag{10b}$$

Substituting Equation (8b) and the derivatives with respect to time of Equations (8a) and (10b) into Equation (3),

$$\frac{\partial \widetilde{T}}{\partial t} + \mathbf{u} \cdot \nabla \widetilde{T} = (1 + \widetilde{T})\left(\nabla \cdot \mathbf{u} + \frac{1}{\gamma p_0}\frac{dp_0}{dt}\right), \tag{11}$$

where the background pressure $p_0(t)$ can be found from Equation (6). The pressure term p(r, t) in Equation (2) may be expressed as

$$p(\mathbf{r}, t) = p_0(t) - \rho_0(t)gz + \widetilde{p}(\mathbf{r}, t), \tag{12}$$

where z is the vertical spatial component and r is the three-dimensional vector spatial coordinates. Combining Equations (2) and (12),

$$\frac{\partial \mathbf{u}}{\partial t} + \frac{1}{2}\nabla|\mathbf{u}|^2 - \mathbf{u} \times \omega + \frac{1}{\rho}\nabla \widetilde{p} - \frac{\rho - \rho_0}{\rho}\mathbf{g} = \frac{1}{\rho}(\nabla \cdot \sigma). \tag{13}$$

The stress tensor σ in Equation (13) can be replaced by the Reynolds stress tensor τ with

$$\tau_{ij} = 2\rho(C_s\Delta)^2 |S_{ij}| S_{ij},$$

where

$$S_{ij} = \frac{1}{2}\left(\frac{\partial u_i}{\partial x_j} + \frac{\partial u_j}{\partial x_i}\right),$$
$$\Delta = (\delta x \, \delta y \, \delta z)^{1/3},$$
$$|S_{ij}| = \sqrt{2S_{ij}\,S_{ij}},$$

where C_s is the Smagorinsky constant.

Lin et al. (2006a) conduct experiments to verify and validate a commercially available CFD code to simulate stratum ventilation. Stratum ventilation is a newly proposed ventilation mode which uses low speed air jets that are strategically placed on the walls of a room at breathing height level to create a layer or stratum of relatively fresh air. Applications of CFD to indoor environment have received great attention in the past two and a half decades. This is mainly fuelled by the availability of relatively cheap yet powerful computers. Numerous authors have commented on the usage of CFD for indoor environment. Chow (1996) gives a comprehensive review on the application of CFD to building service industry. Chen and Zhai (2003) give an assessment of the realism of CFD and its application to indoor environment. They conclude that although CFD is an important tool to compute indoor air problems, a standard procedure must be followed in order to have CFD results being considered accurate. Moreover, the importance of grid resolution for proper results is also discussed. To apply CFD confidently, Oberkampf and Trucano (2002) give an extensive review of current state of verification and validation of CFD in engineering and science. For numerical data verification and validation by experiment, Coleman (1997) gives detailed methodology to assess uncertainties between numerical and experimental results. Srebric and Chen (2002) and Srebric and Zhai (2002) provide step by step account on the procedure for verification, validation and reporting of indoor environment CFD studies. They emphasise the importance of verification and validation of fundamental physical aspects such as basic flow with heat transfer features, turbulence models, auxiliary heat transfer flow models and numerical methods for assessment of predictions. Different types of turbulence models are investigated by Chen (1995) to determine the most appropriate model for indoor air flow computations. He concludes that the re-normalisation group (RNG) k-ε model is the most accurate model for this type of simulation. Indoor air jet ventilation has been studied by researchers for many years. Air distribution in enclosed environments is crucial to thermal comfort and air quality. Computational fluid dynamics (CFD) has played an important role in evaluating and designing various air distributions. Many factors can influence the applications of CFD for studying air distribution. The most critical factors are the selection of an appropriate CFD approach and a turbulence model. Recent advances in CFD approaches and turbulence models provide great potential for improving prediction accuracy of air distribution in enclosed environments. Zhai et al. (2007) summarise recent progress in CFD turbulence modelling and its application to some practical indoor environment studies. Also described are turbulence models that either are commonly used or have been proposed and used recently for indoor environment modelling. Their study further identify a few turbulence models that show great potential for modelling airflows in enclosed environments. Zhang et al. (2007) evaluate the performance of eight turbulence models, potentially suitable for indoor airflow, in terms of accuracy and computing cost. These models cover a wide range of computational fluid dynamics (CFD) approaches. The RNG and a modified model perform the best overall in four cases studied. The other models have superior performance only in particular cases. While each turbulence model has good accuracy in certain flow categories, each flow type favours different turbulence models. Therefore, they recommended both the performance of each particular model in different flows and the best suited turbulence models for each flow category. Equations of laminar and turbulent jet can be found in engineering books such as Schlicting (1979). These equations have been recently applied by Partyka (1995) to determine a mathematical model for air curtain comprising two air jets. Computed results are found to compare well with experimental data. Murakami et al. (1991, 1992) study three-dimensional jets by comparing numerical results obtained from k-ε model with experimental results of a small room, but discrepancies arose. Hence they continue the study by using a different numerical model which is based on algebraic second moment closure scheme and find a better agreement with experimental results. Two and three dimensional jets have been studied by a number of authors. Yue (2000, 2002) investigates the effect of different nozzle sizes for wall jet ventilation. Curd (1986) investigates the behaviour of two and three dimensional wall jets for internal pollutant control. EL-Taher (1983) performs numerical and experimental work to determine behaviour of wall jet. Tornstrom et al. (2001) use different turbulence models to evaluate accuracy of the numerical technique. Just recently, wall confluent ventilation or ceiling air jet is found to perform well in both IAQ and thermal comfort by Cho et al. (2005) and Karimipanah et al. (2005). Side wall air jets have been studied and applied to agricultural industry for ventilation of livestock. Kaiser et al. (1996) investigate various inlets for side wall and ceiling supply ventilation systems. They research into air velocity characteristics of ventilation system where air supply is located at the sides of a full scale experimental livestock building. Zhang and Strom (1999) model drop distance using Archimedes number to control the trajectory of air jets in ventilated air spaces, where drop distance is a function of inlet opening hydraulic diameter and inlet height above the floor. Ge and Tassou (2001) study air jets to create a vertical air curtain for refrigerated display cabinets by finite difference technique.

The re-normalisation group (RNG) k-ε model might be written as

$$\frac{\partial}{\partial t}(\rho\phi) + \frac{\partial}{\partial x_j}(\rho u_j \phi) = \frac{\partial}{\partial x_j}(\Gamma_\phi \frac{\partial \phi}{\partial x_j}) + S_\phi,$$

where

t = time

ρ = air density, kg/m^3

ϕ = 1 for mass continuity

ϕ = u_j (j = 1, 2, and 3) for three components of momentum

ϕ = k for kinetic energy of turbulence

ϕ = ε for dissipation rate of turbulence energy

ϕ = T for temperature

ϕ = c for contaminant concentration

x_j = coordinate

$\Gamma_{\phi,\text{eff}}$ = effective diffusion coefficient

S_ϕ = source term

Different definitions for ϕ, Γ_ϕ, and S_ϕ were given in the following table.

φ	$\Gamma \phi$	$S \phi$
1	0	0
u_j	$\mu + \mu_t$	S_{uj}
k	$(\mu + \mu_t)/\sigma_k$	$G - \rho\varepsilon - G_B$
ε	$(\mu + \mu_t)/\sigma_\varepsilon$	$(C_{\varepsilon 1}G - C_{\varepsilon 2}\rho\varepsilon + C_{\varepsilon 3}G_B)\varepsilon/k + R$
T	$\mu/\sigma_1 + \mu_t/\sigma_\tau$	S_T
c	$(\mu + \mu_t)/\sigma_c$	S_c

where

μ is laminar viscosity,

$$\mu_t = \rho C \mu \frac{k^2}{\varepsilon} \quad \text{was turbulent viscosity,}$$

$$G = \mu_t \frac{\partial u_i}{\partial x_j}\left(\frac{\partial u_i}{\partial x_j} + \frac{\partial u_j}{\partial x_i}\right) \quad \text{was turbulent production,}$$

$$G_B = -g_i\beta\frac{\mu_t}{Pr_t}\frac{\partial T}{\partial x_i} \quad \text{was turbulent production due to buoyancy,}$$

$$R = \frac{C_\mu\eta^3(1-\eta/\eta_0)}{1+\beta\eta^3}\frac{\varepsilon^2}{k} \quad \text{was source term from renormalisation,}$$

$$\eta = S\frac{k}{\varepsilon}, \; S = (2S_{ij}S_{ij})^{1/2}, \; S_{ij} = \frac{1}{2}\left(\frac{\partial u_i}{\partial x_j} + \frac{\partial u_j}{\partial x_i}\right),$$

$C\mu = 0.0845$, $C_{\varepsilon 1} = 1.42$, $C_{\varepsilon 2} = 1.68$, $C_{\varepsilon 3} = 1.0$ were model constants,

$\sigma k = 0.7194$, $\sigma\varepsilon = 0.7194$, $\sigma 1 = 0.71$, $\sigma t = 0.9$, $\sigma c = 1.0$ were Prandtl or Schmidt numbers.

Since the RNG k-ε model was valid for high Reynolds number turbulent flow, wall functions were needed for near wall region where the Reynolds number was low. Low Reynolds number flow investigation uses the following wall functions:

For velocity:

$$U = \left(\frac{\tau}{\rho}\right)^{1/2}\frac{1}{\kappa}\log(\frac{y}{y_*}E),$$

where

U = velocity parallel to the wall

τ = wall shear stress

$\varkappa$ = von Karman constant (=0.41)

y = distance between the first grid node and the wall

E = an integration constant (=9.0)

y^* = a length scale

For kinetic energy of turbulence k:

$$k = \frac{1}{C_\mu^{1/2}} \left(\frac{\tau}{\rho} \right).$$

For dissipation rate of turbulent kinetic energy ε:

$$\varepsilon = \left(\frac{\tau}{\rho} \right)^{3/2} \frac{1}{\varkappa y}.$$

For temperature T:

$$q = h_c (T_w - T) \quad,$$

where

q = heat flux

h_c = convective heat transfer coefficient

T_w = wall temperature

5.2 Thermal Comfort Model

Yigit (1999) develops a two-dimensional computer model by combining two models commonly used for indoor thermal comfort. This two-dimensional model estimates resistance to dry and evaporative heat transfer for a clothing system. Predictions of human responses to thermal comfort using this two-dimensional model are compared with data obtained from laboratory experiments. Burton (1934) is probably the first who published a mathematical model to predict temperature response. Thereafter, an exponential growth of other models has come out. Among these models, only models developed by Fanger (1970,1982) and Gagge et al. (1986) are commonly used. In Fanger model, energy balance at steady state are written as

$$M - W = Q_{sk} + Q_{res} = (Q_c + Q_r + Q_e) + (Q_{cres} + Q_{eres})$$

where M is rate of metabolic heat production in W/m², W is rate of mechanical work accomplished in W/m², Q_{sk} is rate of heat transfer at skin in W, Q_{res} is rate of heat loss from respiration in W, Q_c is convective heat flow in W/m², Q_r is radiative heat flow in W/m², Q_e latent heat transfer rate in W, Q_{cres} is rate of convective heat loss from respiration in W/m², Q_{eres} is rate of evaporative heat loss from respiration in W/m². Total resistance to sensible heat transfer R_t for a fabric of clothing system is calculated as

$$R_t = \frac{T_{sk} - T_a}{Q} A,$$

where T_{sk} is skin temperature in Celsius, T_a is air temperature in Celsius, Q is sensible heat transfer rate in W and A is surface area in m². Total resistance to evaporative heat transfer $R_{e,t}$ provided by a fabric or clothing system and air film is determined by

$$R_{e,t} = \frac{(P_{sk} - P_a)}{Q} A,$$

where P_{sk} is water vapor pressure at the skin in kPa, P_a is water vapour pressure in air in kPa. Thermal resistance of an air layer Ral is given by

$$R_{al} = \frac{1}{h_r + k/t_a},$$

where h_r is linearised radiant heat transfer coefficient in W/m²·°C, t_a is air layer thickness in mm and k is thermal conductivity of air in W/m²·°C. Evaporative resistance of air layer $R_{e,al}$ might be written as

$$R_{e,al} = a[1 - \exp(t_a/b)],$$

where a and b are constant values with a = 0.0334 m²·kPa/W and b = 15 mm. Thermal resistance of outer air layer R_a is

$$R_a = \frac{1}{h_c + h_r},$$

where h_c is convective heat transfer coefficient in W/m²·°C. From experimental measurements by McCullough et al. (1989), h_r and h_c are determined to be 4.2 and 4.9 W/m²·°C respectively. Evaporative resistance of outer air layer $R_{e,a}$ can be determined from

$$R_{e,a} = \frac{1}{h_c LR},$$

where LR is Lewis relation in °C/kPa. In the Gagge model, thermal model that describes two coupled heat balance equations, one applied to each compartment, are

$$S_{cr} = M - W - (Q_{c,res} + Q_{e,res}) - Q_{cr,sk},$$

$$S_{sk} = Q_{cr,sk} - (Q_c + Q_r + Q_e),$$

where S_{cr} is rate of heat storage in core compartment in W/m² and S_{sk} is rate of heat storage in skin compartment in W/m². In terms of thermal capacity and time rate of change of temperature in each compartment

$$S_{cr} = (1 - \alpha)\, mc_{p,b}\, (dT_{cr}/d_\theta)\, /\, A_D,$$

$$S_{sk} = \alpha\, mc_{p,b}\, (dT_{sk}/d_\theta)\, /\, A_D.$$

Evaporative heat loss by regulatory sweating Q_{rsw} is given by

$$Q_{rsw} = \dot{m}_{rsw}\, h_{fg},$$

where $\dot{m}_{rsw}$ is rate at which regulatory sweat is generated in kg/s·m² and h_{fg} was heat of vaporisation of water = 2430 kJ/kg at 30°C. Skin wetness w is determined by

$$w = 0.06 + 0.94\, Q_{rsw}\, /\, Q_{e,max},$$

where $Q_{e,max}$ is maximum of Q_e. Total heat loss is determined by the sum of sensible heat loss Q_{res} and latent heat loss $Q_{e,res}$ to be

$$Q_{res} + Q_{e,res} = [0.0014M(34 - T_a) + 0.0173M(5.87 - P_a)]/A_D.$$

Based on the two node model of Gagge et al. (1986), average core and skin temperatures are known to be

$$T_{sk,n} = 33.7°C,$$

$$T_{cr,n} = 36.8°C.$$

Thermoregulatory control processes (vasomotor, regulation, sweating and shivering) are stimulated and governed by probably five signal triggering processes, namely warm signal from the core ($WSIG_{cr}$), cold signal from the core ($CSIG_{cr}$), warm signal from the skin ($WSIG_{sk}$), cold signal from the skin ($CSIG_{sk}$) and warm signal from the body ($WSIG_b$). Their values were

$$WSIG_{cr} = \begin{cases} 0, & T_{cr} \le T_{cr,n} \\ T_{cr} - T_{cr,n}, & T_{cr} > T_{cr,n} \end{cases}$$

$$CSIG_{cr} = \begin{cases} T_{cr,n} - T_{cr}, & T_{cr} < T_{cr,n} \\ 0, & T_{cr} \ge T_{cr,n} \end{cases}$$

$$WSIG_{sk} = \begin{cases} 0, & T_{sk} \le T_{sk,n} \\ T_{sk} - T_{sk,n}, & T_{sk} > T_{sk,n} \end{cases}$$

$$CSIG_{sk} = \begin{cases} T_{sk,n} - T_{sk}, & T_{sk} < T_{sk,n} \\ 0, & T_{sk} \ge T_{sk,n} \end{cases}$$

$$WSIG_{b} = \begin{cases} 0, & T_{b} \le T_{b,n} \\ T_{b} - T_{b,n}, & T_{b} > T_{b,n} \end{cases}$$

Average temperature of human body T_b is predicted by

$$T_b = {}_\alpha T_{sk} + (1 - {}_\alpha) T_{cr}.$$

Rate of blood flow $\dot{m}_{bl}$ is given by

$$\dot{m}_{bl} = [(6.3 + 200\ WSIG_{cr})/(1 + 0.5\ CSIG_{sk})]/3600.$$

The combined thermal exchange between core and skin can be written as

$$Q_{cr,sk} = (K + c_{p,bl}\ \dot{m}_{bl})(T_{cr} - T_{sk}).$$

Fraction of body mass concentrated in skin compartment α is

$$\alpha = 0.0418 + 0.745/(3600\ \dot{m}_{bl} + 0.585).$$

Rate at which regulatory sweat $\dot{m}_{rsw}$ in kg/s·m^2 is generated by

$$\dot{m}_{rsw} = 4.7 \times 10^{-5}\ WSIG_b \exp(WSIG_{sk} / 10.7).$$

Metabolic energy production due to shivering M_{shiv} is

$$M_{shiv} = 19.4\ CSIG_{sk}\ CSIG_{cr}.$$

Rate of metabolic heat production M in W/m2 is

$$M = M_{shiv} + M_{act},$$

where M_{act} are metabolic heat generated due to activity. Results predicted by the two-dimensional model are in good agreement with experimental data obtained from Tanabe et al. (1994) and Oleson et al. (1988).

5.3 PD, PPD and Mean Age—Parameters for Indoor Environment Assessment

Lin et al. (2005b) apply a validated CFD code to investigate the effect of air supply location under local thermal and flow boundary conditions on the design and performance of displacement ventilation system. Thermal and flow boundary conditions include airflow temperatures, airflow rates, heat sources, contaminant sources, size and geometry of rooms. Results are reported on thermal comfort and indoor air quality (IAQ). Study shows air supply inlet should be located at the centre rather than on the sides of floor. Displacement ventilation was first introduced to Scandinavian countries decades ago. In 1989, it was estimated that displacement ventilation in Nordic countries accounted for 50% market share of industrial applications and 25% of office applications. Applications of displacement ventilation in Hong Kong were started in 1980, however, only a few incidences of installation of displacement ventilation were reported in Hong Kong. For United States, a set of guidelines for designing displacement ventilation were developed. Bauman et al. (1995) and Shute (1995) state that occupants should be 1–1.5 m away from air supply grilles to achieve thermal comfort. Location of air supply on the ceiling according to McCarry (1995) is found to lead to poor circulation at desks in partitioned areas. Gan (1996) uses CFD technique to investigate thermal discomfort by applying comfort equations by Fanger (1989) in the airflow model. He concludes that thermal discomfort can be avoided through an optimisation of air supply velocity and temperature. Experiments performed by Wyon and Sandburg (1990) indicate that thermal comfort is better above table height and thermal discomfort is mostly observed at legs and ankles. Lian (2002) shows that thermal comfort of displacement ventilation system is determined by the distance between occupants and air supply diffusers. A report by European Collaborative Action (ECA 1992) outlines the dangers to health as a result of exposure to certain volatile organic compounds (VOC) in indoor environments. Harmful VOC includes but is not limited to toluene, xylene, formaldehyde, ammonia and benzene. The report also provides recommendations to reduce human exposure to harmful pollutants by specifying ventilation rate for different types of buildings. Thermal comfort is decided by factors such as airflow pattern, temperature distribution, percentage of people dissatisfied due to draft (PD) and predicted percentage of people dissatisfied (PPD). On the other hand, IAQ is decided by factors such as carbon dioxide level, mean age of air and VOC levels. Numerical simulations are done on situations in Hong Kong and are compared with those in United States and Scandinavian countries. Investigation uses the validated code of Lin (2005b) to determine thermal comfort. However, the CFD code only computes airflow pattern and temperature distribution. PD, PPD and mean age of air are not computed from the CFD code. Hence, the model stipulated in ISO 7730:1994(E) developed by Fanger et al. (1989) is applied

PD = $(34 - T)(u - 0.05)^{0.62}$ $(3.14 + 0.37 \times u \times Tu)$ [%],

(For PD > 100%, use PD = 100%),

where

T = air temperature (°C),

u = air velocity (m/s) (For $u < 0.05$ m/s, use $u = 0.05$ m/s),

Tu = turbulent intensity = $100(2k)^{0.5}/u$ [%],

k = turbulent kinetic energy.

The formulae for calculating PPD can also be found in ISO 7730:1994(E)

PPD = $100 - 95 \times \exp(-0.03353 \times PMV4 - 0.2179 \times PMV2)$ [%].

The predicted mean vote, PMV, in the equation is determined by

PMV = $[0.303 \times \exp(-0.036 \times M) + 0.028] \times L$

with

$L = M - W - \{3.96 \times 10^{-8} f_{cl} \times [(T_{cl} + 273)^4 - (T_r + 273)^4] + f_{cl} \times h_c \times (T_{cl} - T) +$
$3.05 \times 10^{-3} \times [5733 - 6.99 \times (M - W) - P_a] + 0.42 (M - W - 58.15) + 1.7 \times 10 -$
$5 \times M \times (5867 - P_a) + 0.0014 \times M \times (34 - T)\}$,

with

M = metabolism (W/m²),

W = external work (W/m²),

f_{cl} = cloth factor,

T = local air temperature (°C),

T_{cl} = cloth temperature (°C),

Tr = mean radiant temperature (°C),

h_c = convective heat transfer coefficient between cloth and air (W/m2·K),

P_a = partial water vapor pressure (Pa).

f_{cl}, T_{cl}, and hc in the above equations are determined by:

f_{cl} = $1.05 + 0.645 \times Icl$ for Icl ≥ 0.078

f_{cl} = $1.00 + 1.290 \times Icl$ for Icl < 0.078

T_{cl} = $35.7 - 0.028 \times (M - W) - I_{cl} \times \{3.96 \times 10^{-8} \times f_{cl} \times [(T_{cl} + 273)^4 -$
   $(T_r + 273)4] + f_{cl} \times h_c \times (T_{cl} - T)$

The convective heat transfer coefficient, hc, is determined from:

$h_c = 2.38 \times (T_{cl} - T)^{0.25}$ for $2.38 \times (T_{cl} - T)^{0.25} \geq 12.1 \times u^{0.5}$

$h_c = 12.1 \times u^{0.5}$ for $2.38 \times (T_{cl} - T)^{0.25} < 12.1 \times u^{0.5}$

where

I_{cl} = clothing insulation (°C·m²/W),

u = air velocity (m/s).

Based on these four parameters, airflow pattern, temperature distribution, PD and PPD, conclusions are drawn for thermal comfort. Mean age of air is defined to be the average time of all air molecules to travel from supply to exhaust. The youngest air is likely to be near the supply and the oldest air is likely to be at stagnant zones or near the exhaust (Grieve, 1989). Li and Jiang (1996) show that mean age of air is governed by the transport equation

$$\frac{\partial}{\partial t}(\rho\tau) + \frac{\partial}{\partial x_j}(\rho u_j \tau) = \frac{\partial}{\partial x_j}(\Gamma_\tau \frac{\partial \tau}{\partial x_j}) + \rho \qquad (14)$$

with boundary conditions:

$\tau = 0$ at the supply diffuser,

$\dfrac{\partial \tau}{\partial x_j} = 0$ at the exhaust and walls.

The ventilation effectiveness, η, is defined as

$$\eta = \frac{c_e - c_s}{c - c_s}, \qquad (15)$$

where

η = ventilation effectiveness,

c_e = contaminant concentration at the exhaust air (ppm),

c_s = contaminant concentration at the supply air (ppm),

c = contaminant concentration in the room air (ppm).

The validated CFD code computes the contaminants concentration distributions. IAQ for both displacement ventilation and mixing ventilation are compared in terms of carbon dioxide concentration level, mean age of air and VOC concentrations of toluene, benzene and formaldehyde. Formaldehyde usually emanates from chairs, tables and wood paneling. Benzene is a major component of paints and resins which has been classified by international agency for cancer research to be carcinogenic to humans. Arp et al. suggest that exposure to benzene has been linked to occurrence of various type of leukemia. Toluene can be found in paints, paint thinners and adhesives.

Lin et al. (2005c) compare the performances of floor-supply displacement ventilation systems with mixing ventilation systems for offices, classrooms, retail shops and industrial workshops under a wide range of Hong Kong thermal and flow boundary conditions, such as a very high cooling load (Figure 3.2). They find that through proper design, displacement ventilation can maintain a thermally comfortable environment that has a low air velocity, a small temperature difference between the head and ankle level, and a low percentage of dissatisfied people.

Figure 3.2 Typical rooms: (a) an office, (b) a classroom, (c) a retail shop, (d) a workshop

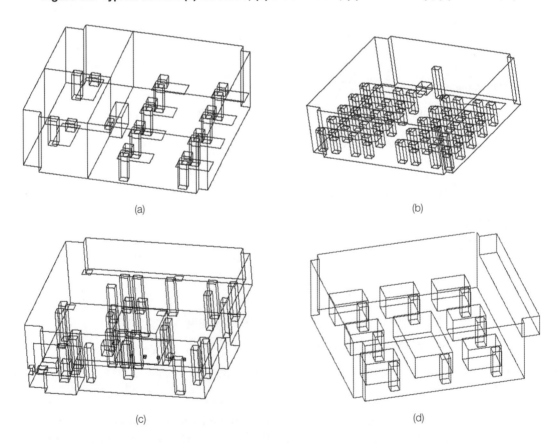

(a)

(b)

(c)

(d)

The Indoor air quality are studied in terms of carbon monoxide, local mean age of air and volatile organic compounds (VOCs) toluene, benzene and formaldehyde which are commonly found in indoor room furnishings (Lin et al. 2005d). Compared with conventional mixing ventilation, displacement ventilation may provide better indoor air quality in the occupied zone.

Lin et al. (2005a) study two cases of air supply locations (Figures 3.3 and 3.4). In the first case, air supplies are located unevenly at one side of a room. Results show that the uneven layout produces two thermal comfort zones. One is located near the air supply, the other one is situated away from the air supply. In the second case, air supply and air exhaust are located at the centre of a room or along the sides of the walls to produce an even distribution of fresh air. In order to maintain an effective supply stream of fresh air to a room, air supply should be placed near the centre of a room.

Figure 3.3 Temperature for retail shop (°C)

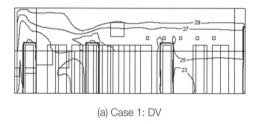 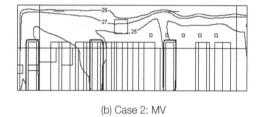

       (a) Case 1: DV   (b) Case 2: MV

Figure 3.4 Benzene concentration for classroom (ppt)

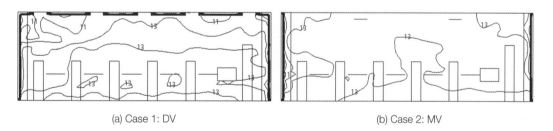

       (a) Case 1: DV   (b) Case 2: MV

The effect of the air supply temperature on the performance of the displacement ventilation system is investigated by Lin et al. (2005e, 2005f). The lower supply temperatures are found to result in higher draft effects, yet increasing temperatures lead to increasing levels of PPD.

Figure 3.5 CO_2 concentration for workshop (ppm)

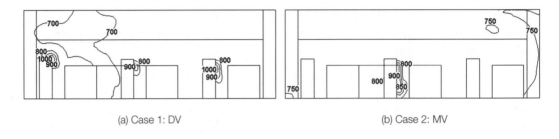

(a) Case 1: DV   (b) Case 2: MV

Figure 3.6 Underfloor supply layouts

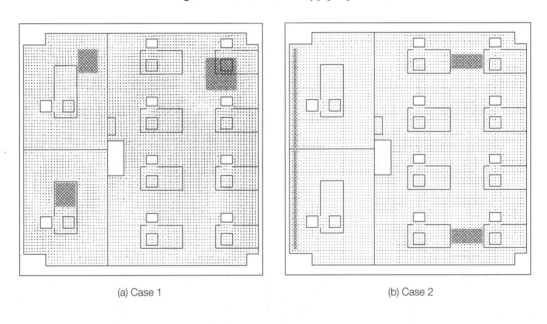

(a) Case 1   (b) Case 2

Figure 3.7 PPD index (%)

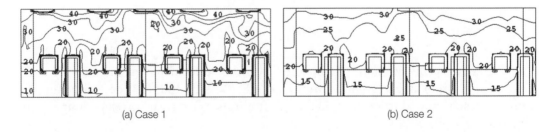

(a) Case 1   (b) Case 2

Figure 3.8 PD index (%)

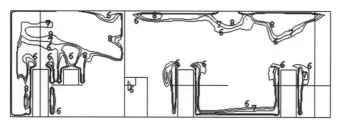

(a) Case 1: supply air temperature = 19°C

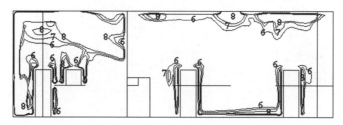

(b) Case 2: supply air temperature = 20°C

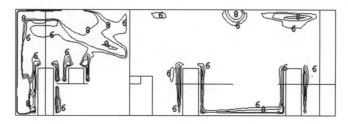

(c) Case 3: supply air temperature = 22°C

The IAQ is measured in terms of carbon dioxide concentration, local mean age of air and three volatile organic compounds (VOCs), namely toluene, benzene and formaldehyde.

Lin et al. (2006b) investigate the effect of the headroom on the displacement ventilation system (DV). It is found that the effect of increased headroom is to improve the thermal comfort of the indoor space.

Figure 3.9 Local mean age of air (s)

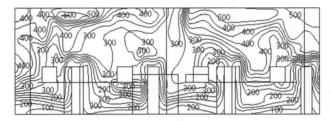

(a) Case 1: supply air temperature = 19°C

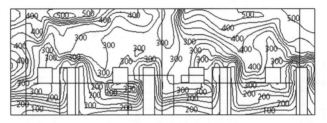

(b) Case 2: supply air temperature = 20°C

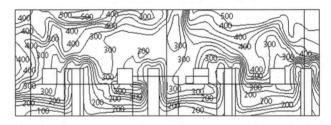

(c) Case 3: supply air temperature = 22°C

Using a validated computational fluid dynamics simulation, Lin et al. (2007) investigate the effect of the position of doors on performance of the displacement ventilation system. It is found that the presence of large heat sources, say, from a window, can cause the lateral movement of airflow and disrupts the convection effect which the displacement ventilation system relies on. Doors can create this situation when they are opened by changing the thermal boundary conditions of indoor spaces. The designer should be made aware of this possibility and make appropriate design decisions to accommodate for this fact.

Figure 3.10 PPD index (%)

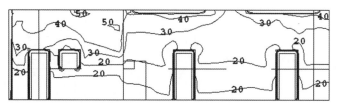

(a) Case 1 (headroom = 2.3 m)

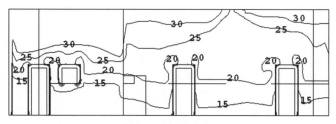

(b) Case 2 (headroom = 2.7 m)

Figure 3.11 Airflow pattern

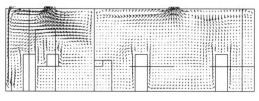 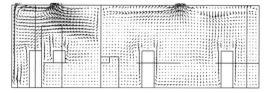

(a) Case 1: doors closed   (b) Case 2: doors opened

Lin et al. (2008) investigate the effect of partitions on the performance of displacement ventilation system. The results indicate that the partitions may significantly affect airflow and performance of a displacement ventilation system. In particular, the presence of a gap above the partition wall is able to improve air distribution owing to less air re-circulation in the upper zone.

Figure 3.12 Temperature distribution (°C)

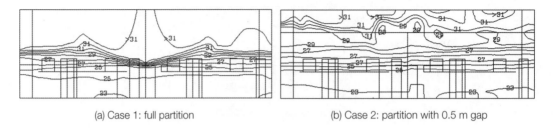

(a) Case 1: full partition   (b) Case 2: partition with 0.5 m gap

Besides indoor environment, the computational fluid dynamics (CFD) technique can also be applied to semi-enclosed space. Lin et al. (2005g) study ventilation effectiveness of the ventilation system within a public transport interchange (PTI) in Hong Kong. A steady state computational model of the PTI is used to investigate and predict the typical pollutant emission pattern for buses. In Hong Kong the displacement ventilation (DV) scheme is often employed for the PTI. The numerical simulation investigates the effectiveness of the DV system in removing pollutants from the occupied zone. An alternative model is proposed where the supply is located at the ceiling and the exhausts are located at the lower part of the columns. It is found that both systems can adequately ventilate the PTI, however the ceiling based air supply system is able to provide improved thermal comfort and indoor air quality (IAQ).

Figure 3.13 A typical PTI in Hong Kong

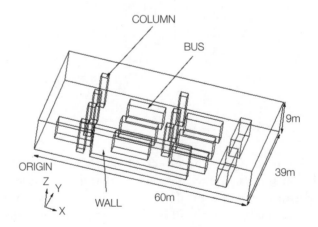

Figure 3.14 NO$_2$ distribution (μg/m^3) at 1.6-meter height

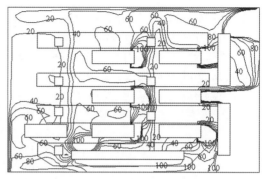

(a) Case 1: displacement system   (b) Case 2: proposed system

6 Conclusion

Office workers in modern society spend most of their time indoors. Nevertheless, indoor air is always a concern which is directly linked to Sick Building Syndromes (SBS) or Building related infection (BRI), which causes eye, nose and throat irritation, dryness of mucous membrane and skin, nose bleeds, skin rash, mental fatigue, headache, cough, hoarseness, wheezing, nausea and dizziness. Due to the inefficiency of currently available ventilation modes, a more effective stratum ventilation mode is developed to facilitate the dilution of indoor air with fresher outdoor air. To study the effectiveness of ventilation systems, room air distribution is simulated by turbulence models. Numerical modelling is shown to be more cost effective than experimentation only if the computational models are well verified and validated. Since turbulence models only account for air velocity and air temperature, PD, PPD and mean age of air models have to be established to account for thermal comfort. Results show that stratum ventilation performs much better than all the previously established ventilation modes in terms of indoor air quality and thermal comfort.

References

Apte, M. G., Fisk, W. J. , & Daisey, J. M. (2000). Associations between indoor CO_2 concentrations and sick building syndrome symptoms in U.S. office buildings: An analysis of the 1994–1996 BASE study data. *Indoor Air, 10*, 246–257.

Arora, S., and Woods, J. E. (1992). Assuring building performance: Redefining professional roles and responsibilities. *Proceedings of the First International Symposium of CIB W82: Future Studies in Construction*, Construction Beyond 2000 (Espoo, Finland): 1–10.

Arp, E. W., Wolf, P. H., & Checkoway, H. (1983). Lymphocytic leukemia and exposure to benzene and other solvents in the rubber industry. *J. Occup Med, 25*, 598–602.

ASHRAE (American Society of Heating, Ventilation and Refrigerating Engineers). (1992). *ANSI/ASHRAE Standards 55–1992: Thermal environmental conditions for human occupancy*. ASHRAE, Atlanta, Georgia, USA.

ASHRAE. (2001) *ASHRAE Standard 62–2001: Ventilation for Acceptable Indoor Air Quality, 1999*. Atlanta, Georgia, USA.

ASHRAE (American Society of Heating, Ventilation and Refrigerating Engineers). (2005). *ASHRAE Handbook: Fundamentals*. SI Edition. Atlanta, Georgia, USA.

Awbi, H. B. (1991). *Ventilation of buildings*. London: E & FN Spon.

Bakó-Biró, Z., Wargocki, P., Weschler, C. J., & Fanger, P. O. (2004). Effects of pollution from personal computers on perceived air quality, SBS symptoms and productivity in offices. *Indoor Air, 14*, 178–187.

Bauman, F. S., Zhang, H., Arens, E. A., & Benton, C. C. (1993). Localized comfort control with a desktop task conditioning system: laboratory and field measurements. *ASHRAE Transactions, 99*(2): 733–749 .

Bauman, F. S., Arens, E. A., Tanabe, S., Zhang, H., & Baharlo, A. (1995). Testing and optimizing the performance of a floor-based task conditioning system. *Energy and Buildings, 22*, 173–86.

Bennetsen, J., Sorensen, J. N., Sogaard, H. T., & Christiansen, P. L. (1996). Numerical simulation of turbulent airflow in a livestock building. *Roomvent '96, Vol. 2*.

Berglund, L. (1989). Comfort criteria in a low humidity environment. *Final Report to Electric Power Research Institute*. Palo Alto, Calif.

Berglund, L. G., Gonzales, R. R., & Gagge, A. P. (1990). Predicting human performance decrement from thermal discomfort and effective temperature. *Proceedings of Indoor Air '90: Fifth International Conference on Indoor Air Quality and Climate (Toronto Ontario) 1*, 215–20 .

Berglund, L. (1995). Comfort criteria in a low humidity standards. *Proceedings of the Pan-Pacific Symposium on Building and Urban Environmental Conditioning in Asia, March 16–18, Nagoya, Japan. Vol. 2. 369–382.*

Brown, S. K. (1999). Assessment of pollutant emissions from dry-process photocopiers. *Indoor Air, 9,* 259–267.

Chan, W. T., Burnett, J., de Dear, R. J., Ng, C. H. (1998). A large-scale survey of thermal comfort in office premises in Hong Kong. *ASHRAE Transactions, 104,* 1172–1180.

Chen, Q. (1995). Comparison of different k-ε models for indoor airflow computations. *Numerical Heat Transfer, 28*(B), 353–369 .

Chen, Q. (1997). Computational fluid dynamics for HVAC: Successes and failures. *ASHRAE Transactions, 103,* 178–187.

Chen, Q., & Chao, N. T. (1996). Prediction of buoyant plume and displacement ventilation with different turbulence models. *Indoor Air, Vol. 1.*

Chen, Q., & Zhai, Z. (2003). How realistic is CFD as a tool for indoor environment design and studies without experiment. *Proceedings of the 4th International Symposium on Heating, Ventilating and Air Conditioning. 62–77.*

Cho, Y., Awbi, H. B., & Karimipanah, T. (2005). Comparison between wall confluent jets and displacement ventilation in aspect of the spreading ratio on the floor. *Proceedings of the 10th International Conference on Indoor Air Quality and Climate, Beijing, 4–9 September,* 3249–3254.

Chow, W. K. (1996). Application of computational fluid dynamics in building services engineering. *Building and Environment, 31*(5): 425–436.

Colemen, H. W., & Stern, F. (1997). Uncertainties and CFD code validation. *Journal of Fluids Engineering, Transactions of the ASE, 119*(4) Dec: 795–803.

Curd, E. F. (1986). Application of two and three dimensional wall jets in the control of air contaminants. *Chemical Engineering Monographs, 24,* 647–662.

Davidson, L., & Nielsen, P. V. (1996). Large eddy simulation of the flow in a three-dimensional ventilated room. *Roomvent '96, Vol. 2.*

Dorgan Associates Inc. (1993). *Productivity and indoor environmental quality study: Final report.* National Energy Management Institute. Madison, Wis.: Dorgan Associates Inc.

Dorgan, C. B., Dorgan, C. E., Kanarek, M. S., & Willman, A. J. (1998). Health and productivity benefits of improved indoor air quality. *ASHRAE Transactions, 104,* 658–666.

EL-Taher, R. M. (1983). Experimental investigation of curvature effects on ventilated wall jets. *AIAA Journal, 21*(11), 1505–1511.

Emmerich, S. J., & McGrattan, K. B. (1998). Application of a large eddy simulation model to study room airflow. *ASHRAE Transactions, 104*, 1128–1137.

Erdmann, C. A., & Apte, M. G. (2004). Mucous membrane and lower respiratory building related symptoms in relation to indoor carbon dioxide concentrations in the 100-building BASE dataset. *Indoor Air, 14*(Suppl 8), 127–134 .

Fang, L., Clausen, G., & Fanger, P. Ole. (2000). Temperature and humidity: Important factors for perception of air quality and for ventilation requirements. *ASHRAE Transactions, 106*, 503–510.

Fanger, P. O. (1970). *Thermal comfort analysis and applications in environmental engineering*. New York: McGraw-Hill.

Fanger, P. O. (1982). *Thermal comfort*. Malabar, Fla.: Robert E. Kriger Publishing Company.

Fanger, P. O., Melikov, A. K., Hanzawa, H., & Ring, J. (1989). Turbulence and draft. *ASHRAE Journal, 31*(7), 18–23 .

Faulkner, C., Fisk, W. J., & Sullivan, D. (1993). Indoor airflow and pollutant removal in a room with desktop ventilation. *ASHRAE Transactions, 99*(2), 750–758.

Faulkner, C., Fisk, W. J., & Sullivan, D. (1995). Indoor airflow and pollutant removal in a room with floor-based task. *Building and Environment, 30*(3), 323–332 .

Federspiel, C. C., Fisk, W. J., Price, P. N., Liu, G., Faulkner, D., Dibartolomeo, D. L., et al., (2004). Worker performance and ventilation in a call center: Analyses of work performance data for registered nurses. *Indoor Air, 14*(Suppl 8), 41–50.

Fountain, M. E., Arens, E., Su, T. F., Bauman, F. S., & Oguru, M. (1999). An investigation of thermal comfort at high humidities. *ASHRAE Trans, 105*, 94–103.

Gagge, A. P., Fobeletes, A. P., & Berglund, L. G. (1986). A standard predictive index of human response to the thermal environment. *ASHRAE Transactions, 92*(2B), 709–731 .

Gan, G. (1996). Numerical investigation of local discomfort in offices with displacement ventilation. *Fuel and Energy, 37*(2), 82–157 .

Ge, Y. T., & Tasou, S. A. (2001). Simulation of the performance of single jet air curtains for vertical refrigerated display curtains. *Applied Thermal Engineering, 21*(2), 201–219.

Goodfellow, H. D. (1985). *Advanced design of ventilation systems for contamination control*. Elsevier.

Grieve, P. W. (1989). *Measuring ventilation using trace-gases*. Denmark: Brüel and Kjær.

Hall, H. I., Leaderer, B. P., Cain, W. S., & Fidler, A. T. (1991). Influence of building-related symptoms on self-reported productivity. *IAQ 91: Healthy Buildings*, pp. 33–35. Atlanta: American Society of Heating, Refrigerating and Air-conditioning Engineers, Inc.

Hetes, R., Moore, M., Northeim, C., & Leovic, K. W. (1995). *Office equipment: Design, indoor air emissions and pollution prevention opportunities*. North Carolina, US Environmental Protection Agency (Report EPA-600/R-95-045).

Hu, S. C., Barber, J. M., & Chuah, Y. K.. (1999). A CFD study for cold air distribution systems. *ASHRAE Transactions, 105*, 614–628 .

Kaiser, K. J., Heber, A. J., Hosni, M. H., & Eakin, G. (1996). Performance of new ceiling and wall ventilation of air inlets. *Applied Engineering in Agriculture, 12*(2), 237–242.

Karimipanah, T., Awbi, H. B., Blomqvist, C., & Sanberg, M. (2005). Effectiveness of confluent jets ventilation system for classrooms. *Proceedings of the 10th International Conference on Indoor Air Quality and Climate, Beijing, 4–9 September 2005*. 3271–3277.

Knebel, D. E., & John, D. E. (1993). Cold air distribution application, and filed evaluation of a nozzle type diffuser. *ASHRAE Transactions, 99*(1), 1337–1348.

Kroner, W. M., & Stark-Martin, J.A. (1994). Environmentall responsive workstations and office-worker productivity. *ASHRAE Transactions, 100*(2), 750–755. Atlanta: American Society of Heating, Refrigerating and Air-conditioning Engineers, Inc.

Kurabuchi, T., Fang. J. B., & Grot, R. A. (1990). A numerical method for calculating indoor airflows using a turbulence model. *NISTIR 89–4211*, National Institute of Standards and Technology.

Launder, B. E., & Spalding, D. B. (1974). The numerical computation of turbulent flows. *Computer methods in applied mechanics and engineering, 3*, 269–289.

Launder, B. E., & Sharma, B. I. (1975). Letters in heat and mass transfer. *Heat and Mass Transfer, Part 1*, 129 .

Leonard, J. J., & McQuitty, J. B. (1986). Archimedes number criteria for the control of cold ventilation air jets. *Canadian Agricultural Engineering, 28*(2), 117–123.

Li, X. & Jiang, Y. (1996). Calculation of age-of-air with velocity field. Paper presented at *Post-IAQ 96 Seminar*, Beijing.

Li, Y., Duan S., Yu, I. T. S., & Wong, T. W. (2004). Multi-zone modeling of probable SARS virus transmission by airflow between flats in Block E, Amoy Gardens. *Indoor Air, 15*, 96–111.

Li, Y., Leung, G. M., Tang, J. W., Yang, X., Chao, C. Y. H., Lin, J. Z., et al. (2007). Role of ventilation in airborne transmission of infectious agents in the built environment—A multidisciplinary systematic review. *Indoor Air, 17, 2*–18 .

Lian, Z. (2002). Experimental study factors that affect thermal comfort in an upward-displacement air conditioned room. *HVAC and R Research, 8*(2), 191–200.

Lin, Z., Chow, T. T., Tsang, C. F., Fong, K. F., & Chan, L. S. (2005a). CFD study on effect of the air supply location on the performance of the displacement ventilation system. *Building and Environment*, 40(8), 1051–1067.

Lin, Z., Chow, T. T., Wang, Q., Fong, K. F., & Chan, L. S. (2005b). Validation of CFD model for research into displacement ventilation. *Architectural Science Review*, 48(4), 305–316 .

Lin, Z., Chow, T. T., Fong, K. F., Wang, Q., & Li, Y. (2005c). Comparison of performances of displacement and mixing ventilations (Part I) — Thermal comfort. *International Journal of Refrigeration*, 28, 276–287.

Lin Z., Chow, T. T., Fong, K. F., Tsang, C. F. & Wang, Q. (2005d). Comparison of performances of displacement and mixing ventilations (Part II) — indoor air quality. *International journal of Refrigeration*, 28: 288–305.

Lin Z., Chow T. T., Tsang, C. F., Fong, K. F., & Chan, L. S. (2005e). Effect of air supply temperature on performance of displacement ventilation (Part I) — Thermal comfort. *Indoor and Built Environment*, 14(2), 103–116.

Lin, Z., Chow, T. T., Tsang, C. F., Chan, L. S., & Fong, K. F. (2005f). Effect of air supply temperature on performance of displacement ventilation (Part II) — Indoor air quality. *Indoor and Built Environment*, 14(2), 117–132 .

Lin, Z., Jiang, F., Chow, T. T., Tsang, C. F., & Lu, W. Z. (2005g). CFD analysis of ventilation effectiveness in a public transport interchange. *Building and Environment*, 41(3), 254–261.

Lin, Z., Chow, T. T., & Tsang, C. F. (2006a). Validation of CFD model for Research into Stratum Ventilation. *International Journal of Ventilation*, 5(3), 345–363.

Lin, Z., Chow, T. T., & Tsang, C. F. (2006b). Effect of headroom on performance of the displacement ventilation system. *Indoor and Built Environment*, 15, 333–346 .

Lin, Z., Chow, T. T., & Tsang, C. F. (2007). Effect of door opening on the performance of displacement ventilation in a typical office building. *Building and Environment*, 42(3), 1335–1347.

Lin, Z., Chow, T. T., Tsang, C. F., Fong, K. F., Chan, L. S., Shum, W. S., et al. (2008).Effect of internal partitions on the performance of displacement ventilation in a typical office environment. *Building and Environment*, in press.

Loomans, M. (1999). Study on the applicability of the desk displacement ventilation concept. *ASHRAE Transactions*, 105, 759–768.

Loudermilk, K. J. (1999). Underfloor air distribution solutions for open office applications. *ASHRAE Transactions*, 105, 605–613.

McCarry, B. (1995). Underfloor air distribution systems: Benefits and when to use the system in building design. *ASHRAE Transactions*, 101(2), 902–11.

McGrattan, K., Rehm, R. G., & Baum, H. R. (1994). Fire-driven flows in enclosures. *Journal of computational Physics, 110*, 285–291.

McGrattan, K. B., Baum, H. R., & Rehm, R. G. (1999). Large eddy simulations of smoke movement. *ASHRAE Transactions, 105*, 426–436.

McQuiston F. C., Parker, J. D., & Spitler, J. D. (2000). *Heating, ventilating, and air conditioning analysis and design.* New York: Wiley & Sons.

Mi, Y. H., Norbäck, D., Tao, J., Mi, Y. L., & Ferm, M. (2006). Current asthma and respiratory symptoms among pupils in Shanghai, China: Influence of building ventilation, nitrogen dioxide, ozone, and formaldehyde in classrooms. *Indoor Air, 16*, 454–464.

Murakami, S. (1988). Visualization of turbulent flow field generated by numerical simulation. *Proceedings of International Symposium on refined Flow Modelling and Turbulence Measurements, Tokyo* .

Murakami, S., Kato, S., & Nakagawa, H. (1991). Numerical prediction of horizontal nonisothermal 3-D jet in room based on the k-ε model. *ASHRAE Transactions, 97*(1), 38–48.

Murakami, S., Kato, S., & Kondo, Y. (1992). Numerical prediction of horizontal nonisothermal 3-D jet in room based on algebraic second moment closure model. *ASHRAE Transactions 98*(1), 951–962.

Nielsen, P. V. (1998). The selection of turbulence models for prediction of room airflow. *ASHRAE Transactions, 104*, 1119–1127.

Oberkampf, W. L, & Trucano, T. G. (2002). Verification and validation in computational fluid dynamics. *Progress in Aerospace Sciences, 38*, 209–272.

Olesen, B. W., Hasebe, Y., & de Dear. R. J. (1988). Clothing insulation asymmetry and thermal comfort. *ASHRAE Transactions 94* (1), 32–51.

Partyka, J. (1995). An analytical design of an air curtain. *International Journal of Modelling and Simulation, 15*(1), 14 .

Preller, L., Zweers, T., Brunekreef, B., & Boleij, J. S. M. (1990). Sick leave due to work-related health complaints among office workers in the Netherlands. *Proceedings of Indoor Air '90, Fifth International Conference on Indoor Air Quality and Climate (Toronto, Ontario), 1*, 227–30.

Raw, G. J., Roys, M. S., & Leaman, A. (1990). Further findings from the office environment survey: Productivity. *Proceedings of Indoor Air '90: Fifth International Conference on Indoor Air Quality and Climate (Toronto, Ontario) 1*, 231–36.

Sakamoto, Y., & Matsuo, Y. (February 1980). Numerical predictions of three-dimensional flow in a ventilated room using turbulence models. *Applied Mathematical Modeling, 4.*

Schlicting, H. (1979). *Boundary Layer Theory.* McGraw Hill, New York .

Sensharma, N. P. (1998). Woods James E and Goodwin Anna K. Relationships between the indoor environment and productivity: A literature review. *ASHRAE Transactions, 104,* 686–701.

Seppänen, O. A., Fisk, W. J., & Mendell, M. J. (1999). Association of ventilation rates and CO_2-concentrations with health and other responses in commercial and institutional buildings. *Indoor Air, 9,* 226–252.

Shaughnessy, R. J., Haverinen-Shaughnessy, U., Nevalainen, A., & Moschandreas, D. (2006). A preliminary study on the association between ventilation rates in classrooms and student performance. *Indoor Air, 16,* 465–468.

Shendell, D. G., Prill, R., Fisk, W. J., Apte, M. G., Blake, D. & Faulkner, D. (2004). Associations between classroom CO_2 concentrations and student attendance in Washington and Idaho. *Indoor Air, 14*(Suppl 7), 333–341.

Shute, R. (1995). Integrated access floor HVAC: Lessons learned. *ASHRAE Transactions, 101*(2), 877–86.

Srebric, J., & Chen, Q. (2002). An example of verification, validation and reporting of indoor environment CFD analyses. *ASHRAE transactions, 108*(2), 185–194.

Srebric, J., & Zhai, Z. (2002). A procedure for verification, validation and reporting of indoor environment CFD analyses. *HVAS & R Research, 8*(2), 201–216.

Sterling, E., & Sterling, T. (1983). The impact of different ventilation levels and fluorescent lightning types on building illness: An experimental study. *Canadian Journal of Public Health, 74,* 385–92.

Sundell, J. (1994). On the association between building ventilation characteristics, some indoor environmental exposures, some allergic manifestations and subjective symptom reports. *Indoor Air,* supplement 2, 1–49. (Thesis)

Sundell, J. (1996). What we know, and don't know, about sick building syndrome, *ASHRAE Journal, 38,* 51–57.

Tanabe, S., Arens, E. A., Bauman, F. S., Zhang, H., and Madsen, T. L. (1994). Evaluating thermal environments by using a thermal manikin with controlled skin surface temperature. *ASHRAE Transactions, 100*(1), 39–48 .

Toftum, J., and Fanger, P. O. (1999). Air humidity requirements for human comfort. *ASHRAE Trans, 105,* 641–647.

Tornstrom, T., Amir, S., and Moshfegh, B. (2001). Flow and thermal characteristics of warm plane air jets (measurements and predictions using different k-ε models). Computational engineering, *Computational Methods and Experimental Measurements, 10,* 33–44 .

Wålinder, R., Norbäck, D., Wieslander, G., Smedje, G., and Erwall, C. Nasal mucosal swelling in relation to low air exchange rate in schools. *Indoor Air, 7,* 198–205 (1997).

Wikipedia. http://en.wikipedia.org/wiki/Amoy_Gardens

Wolkoff, P., Johnson, G. R., Franck, C., Wilhardt, P., and Albrechsten, O. (1992). A study of human reactions to office machines in a climatic chamber. *Journal of Exposure Analysis and Environmental Epidemiology*, Suppl 1, 71–97.

Wyon, D. P., & Sandburg, M. H. (1990). Thermal manikin prediction of discomfort due to displacement ventilation. *ASHRAE Transactions*, 96(1).

Yigit, A. (1999). Combining thermal comfort models. *ASHRAE Transactions, 105*, 149–156.

Yu, I. T. S., Li, Y. G., Wong, T. W., Tam, W., Chan, A., Lee J. H. W., Leung, D. Y. C., and Ho T. (2004). Evidence of airborne transmission of the severe acute respiratory syndrome virus, *New England Journal of Medicine, 350,* 1731–1739.

Yuan, X., Chen, Q., & Glicksman, L. R. (1999a). Performance evaluation and design guidelines for displacement ventilation. *ASHRAE Trans, 105*, 298–309 .

Yue, Z. (2000). Velocity decay in air jets for HVAC applications. *ASHRAE Transactions, 106*(1), 53–59.

Yue, Z. (2002). An experimental investigation of turbulent wall jets. *ASHRAE Transactions, 108*(1), 203–206.

Zhai, Z., Zhang, Z., Zhang, W., & Chen, Q. (2007). Evaluation of Various Turbulence Models in Predicting Airflow and Turbulence in Enclosed Environments by CFD: Part 1 — Summary of Prevalent Turbulence Models. *HVAC&R Research, 13*(6), 853–870.

Zhang, G., & Strom, J. S. (1999). Jet drop models for control of non-isothermal free jets in a side-wall multi-inlet ventilation system. *Transactions of the ASAE, 42*(4), 1121–1126.

Zhang, Z., Zhang, W., Zhai, Z., & Chen, Q. (2007). Evaluation of various turbulence models in predicting airflow and turbulence in enclosed environments by CFD: Part 2—Comparison with experimental data from literature. *HVAC&R Research, 13*(6), 871–886.

Zhivov, A. M., & Rymkevich, A. A. (1998). Comparison of heating and cooling energy consumption by HVAC system with mixing and displacement air distribution for a restaurant dining area in different climates. *ASHRAE Transactions, 104*, 473–484.

Energy Conservation and Central Plant Development

Nowadays, energy conservation becomes a core value in engineering design and management in Hong Kong. In the context of heating, ventilating & air-conditioning (HVAC) and building services, energy conservation opportunities should be properly identified throughout the life cycle of a building project. It is essential to determine the energy saving targets, particularly in the large-scale centralised HVAC and building services systems for the tall buildings, large indoor spaces, or even a district provision.

In this chapter, we will introduce a variety of energy conservation measures that have been advocated by the local government and engineering body. The application potential of the district cooling system in Hong Kong will also be discussed.

Apple Lok Shun CHAN and Square Kwong Fai FONG

Building Energy and Environmental Technology Research Unit
Division of Building Science and Technology
College of Science and Engineering
City University of Hong Kong

1 Energy Conservation Measures in Hong Kong

Energy conservation is one of the common themes of building services design. The depth and coverage depend on the experience of consulting engineering designers and the requirements of clients. Most often, energy conservation equipment and systems are technologically mature and economically viable, therefore incorporating energy saving measures into building services design would be hindered mainly by investment budget and payback period. In order to encourage awareness and wider applications of energy conservation measures for the new and existing buildings, two major energy efficiency schemes have been launched in Hong Kong since 1996 and 1998 respectively—Hong Kong Building Environmental Assessment Method (HK-BEAM) and Hong Kong Energy Efficiency Registration Scheme for Buildings (HKEERSB). Both of the schemes are not statutory, but are undertaken on voluntary basis. However they are welcomed and implemented by various building owners and developers in the last decade. These two schemes are introduced in the following sections; they represent the key energy conservation measures adopted in buildings in Hong Kong.

1.1 Hong Kong Energy Efficiency Registration Scheme for Buildings

As initiated by the Hong Kong SAR government, HKEERSB is a voluntary scheme advocating energy conservation for buildings as strategic promotion of energy efficiency in Hong Kong[1]. This entire scheme has been implemented since 1998, although the code related to the Overall Thermal Transfer Value (OTTV) was promulgated even earlier in 1995. HKEERSB aims to arouse awareness of energy conservation and efficiency in building system designs, provide information on minimum energy performance standards of building, and stimulate the design of energy efficient buildings above the minimum energy performances. HKEERSB applies to commercial buildings and hotels, and it promotes the application of the following five Building Energy Codes[2]:

- Code of Practice for Energy Efficiency of Air Conditioning Installations
- Code of Practice for Energy Efficiency of Lighting Installations
- Code of Practice for Energy Efficiency of Electrical Installations
- Code of Practice for Energy Efficiency of Lift and Escalator Installations
- Performance-based Building Energy Code

Figure 4.1 Building Energy Codes in Hong Kong

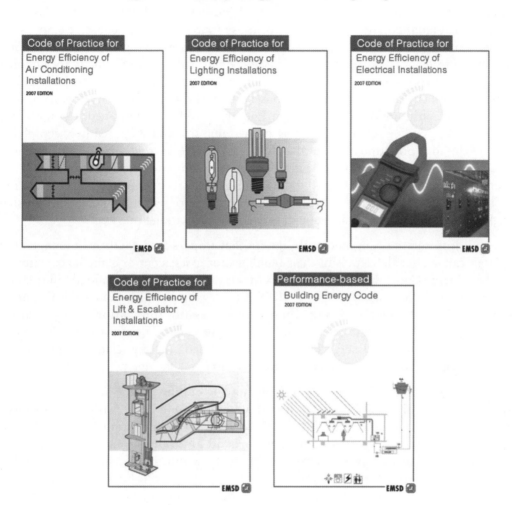

Figure 4.1 shows the 2007 edition of all these five codes. The first four codes are related to energy efficiency of air-conditioning, lighting, lift and escalator, and electrical installations. They are criteria-based and prescriptive, providing information for the minimum design requirements of different building services systems. Focusing on the Code of Practice for Energy Efficiency of Air Conditioning Installations, it encompasses the performance standards for system load design, air side system design, water side system design, automatic control and thermal insulation, and minimum efficiency of air-conditioning equipment. This covers a comprehensive system design from load calculation, equipment and system selection of both centralised and unitary installations.

The fifth one, Performance-based Building Energy Code, specifies the energy efficiency from a holistic approach instead of individual building services. This code has provided an alternative path on top of those four prescriptive codes, it emphasises on the consideration of total energy consumption of various systems, their interrelation and energy trade-off. This performance-based code provides more flexibility for building services installations, as well as room for innovative and renewable energy design. In order to consider the degree of energy conservation of the designed building, its annual total energy consumption, called Design Energy, would be determined. On the other hand, a corresponding reference building of the same size and shape, entirely complying with all the four prescriptive codes, would be developed. Its total energy consumption, called Energy Budget, would then be determined. If Design Energy is less than or equal to Energy Budget, this Code is deemed to be complied. Hourly energy and plant simulation are usually indispensable in this kind of computation, since monthly calculation by using single design day is not accurate enough for such benchmarking purpose.

Building services engineers, designers, developers and facilities management agencies can submit the details of their buildings for assessment of compliance with the Building Energy Codes. HKEERSB requires the certification of the submitted information by a Registered Professional Engineer (RPE) of relevant discipline under the Engineers Registration Ordinance in Hong Kong. For successful application, a registration certificate would be issued to the building meeting the codes. The registered building can also adopt the scheme's "Energy Efficient Building Logo" on any related documents, so as to publicise its achievement of energy conservation and efficiency.

Starting from 2007, the existing commercial buildings and hotels can also be registered under HKEERSB. The existing buildings should demonstrate good energy performance and audited according to "Guidelines on Energy Audit"[3] and certified by RPE. The annual electricity consumption and appropriate energy management opportunities, and energy consumption intensity would be considered. An online "Benchmarking Tool for Buildings"[4] is available to let the building practitioners compare their energy consumption levels with others in the same building type, set future targets, identify measures to reduce energy consumption, and to achieve HKEERSB.

In Hong Kong, many famous building projects have already been registered under HKEERSB, including Two International Finance Centre, Hong Kong Science Park, Cyberport, Hong Kong International Airport Passenger Terminal Building, The Center, Cheung Kong Centre, Lippo Sun Plaza, International Wetland Park and Visitor Centre, and a number of government office buildings, schools, hospitals and quarters.

HKEERSB demonstrates the commitment of the Hong Kong government in energy conservation and efficiency for buildings. The minimum performance requirements for different energy consuming systems in buildings would provide the baseline below where the energy inefficient designs and installations are clearly defined. Both the criteria- and performance-based codes formulate a comprehensive series of strategies for the building designers and owners in the scope of energy conservation.

1.2 Hong Kong Building Environmental Assessment Method

HK-BEAM is another voluntary assessment scheme for energy conservation and efficiency; it has been launched since 1996, earlier than the governmental HKEERSB. Unlike HKEERSB, HK-BEAM was initiated by the private sector, the Real Estate Developers Association of Hong Kong, and it is now owned and operated by HK-BEAM Society. HK-BEAM assesses the building projects in a wide spectrum including the building site, construction materials, energy consumption, water quality and consumption, indoor environment and innovative installations from the perspective of the construction and operational lifetime. The primary aims of HK-BEAM are to advocate the construction of sustainable buildings in Hong Kong, give recognition to buildings with improved performance, provide building practitioners and professionals with a common set of performance standards, and to reduce the environmental impacts of buildings throughout the life cycle. Therefore the assessment criteria of HK-BEAM are not limited to the aspect of energy conservation, it actually adopts a holistic building environmental evaluation approach. The standard and supporting process of HK-BEAM has already been applied for both new and existing buildings. The types of buildings include commercial, residential, institutional buildings and mixed use complexes. It would be a useful means to improve and benchmark the performance in planning, design, construction, commissioning, operation and management of the entire buildings.

The HK-BEAM standards have been continually evolved. The latest standards have been issued since 2004:

- HK-BEAM 4/04 "New Buildings" An environmental assessment for new buildings version 4/04; and

- HK-BEAM 5/04 "Existing Buildings" An environmental assessment for existing buildings version 5/04.

These standards are available from the official web site of HK-BEAM Society[5], a simplified Chinese version of the first standard is also available there.

Figure 4.2 HK-BEAM standards

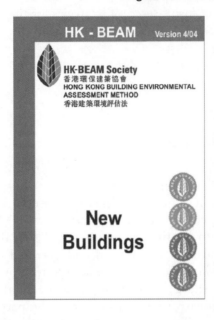 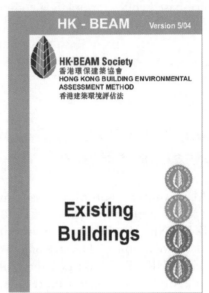

The certified buildings through HK-BEAM would be labelled as Platinum, Gold, Silver or Bronze according to both the overall applicable credits gained and the importance of Indoor Environmental Quality (IEQ) shown in Table 4.1:

Table 4.1 Award classification of HK-BEAM

	Overall Credits	Minimum IEQ
Platinum (excellent)	75%	65%
Gold (very good)	65%	55%
Silver (good)	55%	50%
Bronze (above average)	40%	45%

Many famous existing and new building projects in Hong Kong have been awarded the HK-BEAM labels[6], such as Two International Finance Centre, One Peking, Bank of China Tower, HKBC Headquarters, Taikoo Place, Exchange Square, Jardine House, Hong Kong Science Park, Cathay Pacific City, Festival Walk, Cyberport, CityU Student Hostels Phase III, Ngong Ping 360 Cable Car Project, etc.

In the aspect of energy use and conservation, HK-BEAM advocates the provision of energy conservation in the detailed design of building services systems. Credits are awarded on the basis of enhanced energy efficiency and performance, installation of energy efficient systems and equipment, and provision for energy management. HK-BEAM covers the following four major areas in energy efficiency and conservation:

a. Annual energy use—this considers the energy use in different types of buildings, including the commercial, hotel, educational, residential, mechanically ventilated (MV) buildings.

b. Energy efficiency systems—this assesses the embodied energy in buildings, structural elements, ventilation systems, lighting systems, hot water supply systems, lift and escalator systems, electrical systems, and renewable energy systems.

c. Energy efficient equipment—this focuses the air-conditioning units, clothes drying facilities, energy efficient lighting in public areas, heat reclaim, mechanical ventilation in hotel buildings, and energy efficient appliances.

d. Energy management—this relates to testing and commissioning, operation and maintenance, metering and monitoring.

As compared with HKEERSB, HK-BEAM has the following features in energy conservation:

- HK-BEAM is applied to a variety of building types, while HKEERSB is only provided for the commercial buildings and hotels at this stage.

- OTTV is not only applied to commercial buildings and hotels, but also to other types of buildings under the scheme of HK-BEAM.

- For heating, ventilating and air-conditioning (HVAC) systems, the predicted annual air-conditioning electricity use is already determined by the characteristics of the holistic building and air-conditioning system, such as the design indoor conditions, coefficient of performance of chillers, efficiency of water side and air side equipment.

- For lighting systems, the interior lighting control is a pre-requisite for credits, and the maximum allowable lighting power density is assessed.

- For the energy efficiency of electrical installations, lifts and escalators, credits are awarded if the respective Codes of Practice for Energy Efficiency are complied.

As a whole, HK-BEAM advocates its performance-based standards and encourages the pursuit of energy conservation and efficiency, together with other sustainable designs and measures for buildings. It suggests a comprehensive building environmental assessment for both the developers and designers to improve the buildings, no matter existing or new, to a whole-life and more sustainable performance. HK-BEAM drives to provide a healthier, efficient and environmentally friendly working and living environments in Hong Kong.

1.3 Energy Audit and Energy Management

As discussed in the previous sections, both HKEERSB and HK-BEAM are energy conservation schemes for both new and existing buildings. To promote energy conservation and efficiency for existing buildings is a challenge to the building owners and facilities management practitioners. Suitable incentives should be available in order to encourage the implementation of energy conservation schemes for existing buildings. At this moment, there are no tangible incentives offered from the local government. However through careful identification of the energy conservation opportunities, economic advantages in energy and operating costs can be achieved. This is the spirit of energy audit, an investigation of energy consuming installations to ensure that energy saving can be achieved without sacrificing the operation requirements and indoor comfortable environment. Energy audit is similar to financial accounting, but is used for energy accounting.

To implement energy audit successfully, top management's commitment and initiative is essential, so that sufficient resources would be allocated for such purpose. Energy audit is an effective energy management tool; it is more than money saving. Its ultimate goal is efficient energy use and effective energy saving, which implies environment protection and sustainable development. In energy audit, the key exercise is to identify the possible energy management opportunities (EMOs) of the existing building after a series of information collection, measurement and data analysis works. There are generally three categories, as classified according to the need of investment cost:

- EMO Cat I: No investment cost is practically required in this category. Usually better settings of installed systems, schedules of equipment and house keeping are the possible opportunities.

- EMO Cat II: Low investment cost is needed and short payback period would be involved in this category. This may involve addition of automatic control systems, replacement by energy efficient components, or minor modification of systems.

- EMO Cat III: Relatively high investment cost is incurred and reasonable payback period would be achieved. Substantial addition or modification is required for certain part of equipment or systems.

In the context of HVAC services, it is not difficult to find out the three categories of EMOs. EMO Cat I may be the temperature reset of air side and water side equipment, for instance, the supply air temperature of air handling units, chilled water supply temperature of chillers. Switching off the unused air-conditioning units and suitable setting of room thermostats is an easy way of better house keeping in this category. EMO Cat II may be the addition or modification of automatic control equipment, such as to install timers for better system scheduling, replace conventional controllers with the PID ones for faster and more accurate control, and to add more sensing probes for better monitoring the operating performance. EMO Cat III may be commonly related to the installation of energy conserving equipment, like the variable speed drives for pumps or fans, enthalpy wheels for heat recovery between outdoor and exhaust air streams, or even replacement of air-cooled chillers by water-cooled ones and evaporative cooling towers. Since it would involve much disruption to general building operation from EMO Cat III, these modification works would be commonly scheduled in the cold season.

Commitment of top management and allocation of financial budget is crucial for both EMO Cats II and III. The payback period is commonly three to four years from commercial viewpoint, it depends on the requirements of technical renovation. In order to conduct a comprehensive energy audit and present the recommendations to the top management, a systematic way to carry out energy audit is essential. Generally there are eight steps of energy audit as follows:

i. Defining the scope of energy audit

ii. Forming the energy audit team

iii. Estimating time frame and budget

iv. Collecting the building information

v. Conducting the site inspection and measurement

vi. Analyzing the collected data

vii. Identifying the EMOs

viii. Preparing the energy audit report

The top management of the building would base on the recommendations in the energy audit report to decide the priority and necessity of investment, hence the allocation of financial budget. If a building complex is involved, the energy audit may be conducted by zones or phases; this allows the top management to consider the cash flow and return period more readily.

2 Development of Central Refrigeration Plant

In a central refrigeration plant, chiller can be viewed as the "heart" from which cold refrigerant is produced. Heat transfer between the refrigerant and chilled water takes place in the evaporator of a chiller. The "blood" will be in the format of chilled water and is supplied to various terminal units in a building for conditioning the indoor air.

2.1 Vapour Compression Chiller and Absorption Chiller

There are different types of chillers available in the market. The most common one is centrifugal chiller operating under vapour compression cycle in which four major components are included, namely evaporator, compressor, condenser and throttling device. The evaporator is actually a shell-and-tube type heat exchanger in which cold liquid refrigerant absorbs the heat energy from the chilled water returned from the terminal units. After having absorbed the heat energy, the refrigerant is vapourised and then enters the next major component: compressor. The function of a compressor is to compress the vapour refrigerant in order to raise its pressure and temperature. The hot vapour refrigerant is then discharged to the condenser, which is also a shell-and-tube heat exchanger, for rejecting the heat energy to a heat sink. The heat sink can be outdoor air, sea water or river (details will be addressed in the next section). In the condenser, the vapour refrigerant is condensed as liquid and then passes into a throttling device which can be a float valve or a series of orifice plates. Both the pressure and temperature of the liquid refrigerant are reduced in this device, and a mixture of liquid and vapour refrigerant is formed and then fed into the evaporator. This is the basic refrigeration cycle performed by the four major components of a centrifugal chiller, for which inlet vanes are widely used as a stepless capacity control. Variable speed drive, direct diesel engine or steam turbine drive can also provide capacity control by adjusting the speed of the centrifugal compressor.

CFCs (chlorofluorocarbon) and HCFCs (hydrochlorofluorocarbon) were used in the past as refrigerants for centrifugal chillers. On the international level, there is increased emphasis that chlorine based refrigerants, especially CFCs (e.g. R-11 and R-12) are the cause of ozone layer depletion and greenhouse effect of the earth, from which adverse impact on human health and the environment are resulted. With the international agreement, The Montreal Protocol, the use and emission of CFCs and HCFCs will be eliminated. Environmentally safe refrigerant such as HFC-134a and ammonia (R-717) should be considered as the first-choice refrigerant for centrifugal chillers.

Another major type of chiller is absorption chiller in which water is used as the refrigerant. Heat energy is used to produce the refrigeration effect, without the use of any prime mover, in an absorption system. During the refrigeration cycle, aqueous lithium bromide (LiBr) is used to absorb and carry the refrigerant (water vapour). Heat is removed from the cooling medium during the evaporation of liquid refrigerant. Then the vapour refrigerant is absorbed by the LiBr solution. Heat energy is supplied to the solution to boil off the vapour refrigerant from the absorbent to re-concentrate the solution. After that, the vapour refrigerant is condensed into liquid form by condenser water.

The coefficients of performance (COP) of absorption chillers range from 0.6 to about 1 which are not compatible with the energy performance of the electricity-driven centrifugal chillers. However, since absorption chiller uses water as the refrigerant, it is more environmental friendly due to the zero ozone depletion potential. Moreover, as absorption chiller can use gas as the fuel to drive the refrigeration cycle, the installation of absorption chiller for use during on-peak hours would be economically beneficial when the cost of natural gas is lower than the electricity cost.

2.2 Heat Rejection System

In a central refrigeration plant, the refrigeration cycle cannot work without a heat rejection system. It is a system for extracting condensing heat from the hot refrigerant in the condenser of a chiller and then to reject to the outdoor space, either into air or water. There are four major types of heat rejection systems commonly used in vapour compression refrigeration system, namely air-cooled condenser, water-cooled condenser using underground or sea water directly, water-cooled condenser with cooling tower and evaporative condenser.

Air-cooled condenser consists of a coil, fan, motor and casing. In an air-cooled condenser, vapour refrigerant is condensed by means of a transfer of heat energy to air passed over the coil. The condenser coil is constructed with copper tubes and aluminum fins. For providing uniform air stream through the condensing coil, the fan is usually located downstream of the coil for better heat transfer. Air-cooled condenser is mainly used in medium and small sized application or were the water is scarce.

For central refrigeration plant located near the sea front, water-cooled condenser using sea water can be used. Sea water is pumped into a plate or shell-and-tube heat exchanger by centrifugal pump. Condensing water enters the heat exchanger and is cooled by the sea water. After having absorbed heat energy, the sea water is discharged to the sea. The function of the heat exchanger is to protect the water-cooled condenser from the corrosive effect of sea water. If underground or river water is used as the condensing water, no heat exchanger is needed and the water can be pumped through the water-cooled condenser directly. The application of water-cooled condenser is suitable for large scaled building when steady supply of underground or sea water is available.

For water-cooled condenser with cooling tower, recirculating condenser water from the condenser of a chiller is evaporatively cooled by contact with atmosphere. In a cooling tower, the major components are water distribution system, fill, fan, motor casing, basin and sump. The condensing water coming from the condenser is uniformly sprayed over the fill. Atmospheric air is drawn by a fan through an intake louver and goes across the fill and comes in contact with the water film. The evaporatively cooled water falls into the water basin of the cooling tower and flows back to the condenser. Water-cooled condenser with cooling tower is appropriate for medium and large scaled projects. During the design and operation stages, attention has to be paid for the prevention of Legionnaires' Disease.

In evaporative condenser, heat rejection is taken place through removing the latent heat of condensation of hot refrigerant by the evaporation of water spray. It consists of a condensing coil, fan and motor, water distribution system, sump, recirculating pump and casing. Refrigerant gas is condensed by means of a combination of sensible and latent heat transfer process. Rejected heat is dissipated by water sprayed over the coil surface. It is then transferred to the air passing over the coil. This type of condenser is mainly used in industrial application.

In Hong Kong, air-cooled chillers and water-cooled chillers with cooling tower are commonly used in most commercial buildings. Indirect seawater-cooled chillers are employed in some buildings which are located near the sea front. Absorption chillers may be used in hospitals where excessive steam can be supplied from the boilers.

2.3 Chilled Water System

For distributing the chilled water from the chiller plant to the terminal units on various floors of a building, a closed-loop water circuit is constructed. There are two major types of chilled water system in the water circuit, namely differential pressure bypass (DPB) system and automatic staging variable flow chilled water system.

The differential pressure bypass setup is applied to chilled water system where constant chilled water pumps are designed and two-way valves are provided for the terminal units. In the DPB circuit, a differential pressure bypass setup is installed to sense the differential pressure between main pipes of chilled water supply and return. A two-way motorised modulating valve is installed for bypassing the chilled water from the supply chilled water pipe to the return to maintain the designed differential pressure. As the cooling demand is reduced, the control valves of some terminal units are closed or partly closed; system pressure will be built up and sensed by the DPB setup. Therefore the DPB controller will open the modulating valve to bypass excessive chilled water so that the original designed differential pressure can be restored. During full load condition, the valve is fully closed, and no chilled water will flow through the bypass pipe.

In automatic staging variable flow chilled water system, the primary and secondary circuits are interconnected by a short common pipe with extremely low pressure drop. In the primary circuit, there are chillers with associated primary water pumps which maintain constant flow in the circuit. In the secondary circuit, there is more than one distribution pump and many terminal units. Two-way valves modulate the flow rate of chilled water through the terminal units during part load operation. A common pipe, sometimes called the decoupler, serves as a bypass line between the primary circuit and secondary circuit. A flow meter and a flow switch are mounted on the bypass line for measuring the water flow rate along the bypass pipe. At the design load, the volume flow rate in the primary circuit equals to that in the secondary circuit. During part load condition, surplus chilled water from the chillers in the primary circuit will bypass the secondary circuit and return to the chillers directly through the bypass pipe. When the flow rate of the surplus chilled water is 115 percent of that of the primary chilled water pump, a chiller and its associated pump will be switched off to cater for the reduced cooling demand. The reason for using 115 percent instead of 100 percent is to avoid hunting situation. When the cooling demand is increased again, the chilled water supply from the primary circuit is less than the chilled water flow rate in the secondary circuit. A portion of the chilled water from the terminal units will flow through the bypass pipe in an opposite direction indicating a deficit of chilled water supply. A chiller and its associated pump will then be started again. This system can provide variable chilled water flow at the secondary circuit with separate secondary pumps and constant chilled water flow through the evaporator of the chiller. Thus pumping energy can be saved during part load operating condition. According to the calculation by Rishel[7], the annual pumping energy consumption in this system is about 35% of that consumed by a primary-secondary system using constant flow water pumps with three-way control valves.

2.4 District Cooling System

With the rapid development of densely populated urban areas, the highly packed building clusters with high thermal loads facilitate the development of a new form of central air-conditioning system namely District Cooling System (DCS). DCS is a massive cooling energy production scheme in which chilled water is produced in a remote central chiller plant, and is delivered to serve a group of consumer buildings through a closed-loop piping network. The system efficiency of a DCS is overall higher than the individual chiller plant installed at a single building. This is achieved, firstly, through the mass-scale production in which larger chillers with higher efficiency are in use, and secondly, through the thermal load diversion in which the installed cooling capacity at the central chiller plant can be smaller than the total capacity to be installed at the customer buildings. Moreover, DCS customers can utilise their building space more effectively since the installation of their own cooling facilities is no longer required. From the environmental point of view, the pollutant emissions and wastes (like CFC) from a remote district cooling plant site are easier to be taken care of than those released from small and scattered cooling plants all over the district. The above economical and environmental benefits are best experienced in modern city where the cooling load density is high—typically in association with tall buildings.

A DCS scheme typically consists of groups of parallel chillers, heat exchangers, parallel or series pumps, complex pipe work, in association with control valves, temperature or flow sensors and controllers. Chilled water is circulated between the DCS and the buildings to be served via a decoupler chilled water piping system comprising a primary circuit and a secondary circuit in association with a heat exchanger in each building zone. Each chiller is equipped with its associated primary pump, and the two are interlocked together. The primary pumps are all constant speed and constant flow. The secondary distribution pumps are responsible for overcoming the pressure losses incurred by chilled water flow between the DCS and the buildings. The speed of secondary pump has to be adjusted to maintain a stable pressure differential between the water supply and return. The secondary circuit is hydraulically de-coupled from the primary circuit by the presence of a decoupler bypass pipe between the two circuits. The secondary pumps are all variable speed and variable flow. A control valve will be provided at each building branch for regulating the chilled water flow rate to be fed to each building zone. Cooling energy exchange between the secondary circuit and the local chilled water loop of each building zone is via a heat exchanger. Flow restriction imposed by the control valve operation made the overall friction loss from the central plant to each heat exchanger basically the same. The mass flow rate of the circulating chilled water is minimised by widening the difference of the supply and return chilled water temperatures. The chilled water pipe work is insulated to minimise distribution energy losses.

There are many DCS installations found in different countries of the world, such as Europe, USA, Japan and other Asian countries. In Europe, the application of DCS is increasing steadily, covering 14 countries (including France, Germany, UK, Sweden, Norway, Portugal, Italy, etc.) with about 70 cooling plants[8]. Although the climate in those European countries is colder, district cooling is well developed and has been so for a long time. In USA, the first commercial DCS has been in operation since 1962 at Hartford with a cooling capacity of 52 MW[9]. Since this project has been established, district energy utilities grew substantially. Now, there are over 6,000 district heating and cooling systems in USA which provide 360,000 MWh of energy (in many applications, DCS is integrated with district heating plant to form a District Heating and Cooling (DHC) system). Most of the output is used in institutional systems which serve groups of buildings owned by one entity, such as college, university, hospital or military base.

Japan is one of the pioneers in the field of DCS. The first application was constructed at the 1970 EXPO in Osaka of capacity 130 MW. The growth in number of DHC locations increases significantly over the past 30 years. Currently, there are more than 220 DHC installations which mainly serve office buildings, where air-conditioning is needed, primarily for thermal comfort and computer operations. In other Asian countries such as Malaysia, Korea and China, the market of DCS continues to grow with the climatic change and increased standard of living. The major driving forces of developing DCS systems around the world are air pollution reduction and energy saving. It is recognised that DCS has created a win-win situation for building owners, users and the society as a whole.

3 Applications in HVAC Design and Energy Management

3.1 Effective Methods for HVAC Optimisation Problems

A number of optimisation methods have been applied in handling HVAC optimisation problems. There has been continuous development of optimisation methods in last two decades and such problems can be effectively tackled according to their nature and the optimal solution identified. These methods have their own features as well as limitations in applications.

3.1.1 Analytical approach

Differentiation is a classical optimisation method to determine the optimum for a continuous objective function. This is efficient for a smooth function with a single problem variable. Once the number of problem variables is up to two or more, the roots from the partial derivatives produce families of lines, and the optima can be determined from the intersection of these lines. The global optimum can be identified from a group of optima in an effective way. If equality constraint functions are involved, the method of Lagrange multipliers can be applied to include the constraint functions into the objective function. The roots of the partial derivatives of this modified objective function still produce an equation set to determine the optimum. To handle inequality constraints, usually the Kuhn-Tucker (KT) method would be adopted. However the solution of the KT equations is just the necessary condition for optimality, so the involvement of constraint functions may impede the search reliability of optimum. The obvious limitation of the analytical approach is its incapability to handle optimisation problems with discrete variables, or those without first and second derivatives. It has problems with multimodal optimisation since it may be easily trapped at a local optimum. It also has problems in handling inequality constraints.

3.1.2   Gradient-based and nonlinear programming methods

The method of steepest descent and the Newton method are classical gradient-based numerical methods, and they have been in use for more than half a century. The method of steepest descent starts at an arbitrary point on the search space and searches along the direction of the steepest gradient. The new gradient is orthogonal to the previous one, so that the search is gradually directed towards the optimum. As compared to the steepest descent method, Newton method has a significant improvement on searching efficiency. This is achieved by working on an approximation from multidimensional Taylor series expansion of the objective function and the corresponding matrix of second derivatives called Hessian (Jacobian) matrix. However the Hessian matrix of Newton method may be difficult to determine, and the search of optimum may be unstable and possibly erroneous in multimodal problems. As a result, some quasi-Newton techniques have been developed to replace the Hessian matrix by an approximation one, but necessary checking and remedial works are included in order to provide a trustworthy solution. The sequential quadratic programming (SQP) algorithm is a generalisation of the Newton method for constrained optimisation. It would find a step away from the current point by minimizing a quadratic model of the problem. In each step of iteration, a quadratic approximation is developed for the Hessian matrix of the Lagrangian function. This is then used to generate a quadratic programming subproblem whose solution is used to form a search direction. However these gradient-based methods, similar to those of the analytical approach, are limited to continuous and unimodal functions with the existence of the first and second derivatives.

3.1.3 Exhaustive search

The exhaustive search method does not need any derivative information of objective and constraint functions, but it demands a considerable computational time, particularly if the search interval for each variable is divided too small. If two problem variables are involved, a coarse topographical landscape can be developed first. Then the possible location of the global optimum can be identified, and the exhaustive search can be focused on the potential region with fine resolution. However if the problem is multidimensional and has more than two problem variables, the topographical preview cannot be implemented. On the other hand, the degree of resolution may be based on the optimisation experience or the bounds of the problem variables. If the resolution is still relatively coarse with respect to the problem, the topographical landscape may even mislead the location of the global optimum.

3.1.4 Dynamic programming

Dynamic programming is a method for solving an optimisation problem by memorising the sub-problem solutions rather than recomputing them. This is useful to tackle optimal supervisory control and sequential decision problems. The three main steps of dynamic programming are initialisation, matrix-fill and trace-back. In general, dynamic programming has a simple and logical nature. It works well to find the optimum for problems that can be formulated according to the structure of dynamic programming. However not all kinds of HVAC optimisation problems can be transformed into the required format of this method. The memory-demanding feature of dynamic programming may be another limitation if the dimension of problem variables and their operating ranges are large.

3.1.5 Direct search

The direct search is another method that information about derivatives is not required. This has the attractive property of working with a population of points as well. This is a significant improvement over the exhaustive search in terms of optimisation efficiency. The development of the direct search method is exemplified by the pattern search algorithm of Hooke and Jeeves[10]. The search starts from an initial feasible point in the search space. A series of exploratory probes that form a searching mesh are made along the axes of problem variables. The objective function value of each probe is evaluated, recorded and compared, so that it can inform an accelerated move to a new base point for another set of exploratory probes. This process continues until the optimum is determined. Another common direct search method is the complex method[11]. Its principle is to explore the optimum by rejecting infeasible points. Firstly a group of trial points called simplex are generated randomly within the feasible space. All the points in the simplex are checked for their feasibility against the constraint functions. If any trial point has constraint violation, it would be successively moved halfway back towards the centroid of the existing points of the simplex until it is feasible. Therefore the infeasible point would not be maintained, but relocated within the feasible space. The direct search methods have been applied for handling the HVAC optimised design problem[12]. But these methods have difficulties with constrained problems where the solution lies on the boundary of any constraint functions.

3.1.6   Metaheuristic optimisation methods

In the recent two decades, a number of heuristic methods that do not require the derivative information have been advocated. The common ones are simulated annealing[13], tabu search[14], ant colony optimisation[15] and particle swarm optimisation[16]. Except the simulated annealing used by Koeppel et al.[17], the other heuristic methods have not been applied in HVAC problems. Each method has its own features and suitability in application, but all of them depend on problem-specific parameters, which are not generally transferable to other problems.

3.1.7   Evolutionary algorithm

Evolutionary algorithm (EA) is a probabilistic and population-based heuristic algorithm developed from the Darwinian paradigm of evolution, which is often viewed as analogous to optimal exploration and optimisation. The essential steps are derived from the fundamental principles of variation and selection of the Darwinian evolution throughout generations. Selection is to implement "survival for fittest". The fitness of offspring is evaluated first, and the highly fit offspring would be selected to survive in next generation. The fitness of an individual refers to the performance of adaptation to the environment for survival according to the Darwinian paradigm of evolution. In typical optimisation problem, it is the evaluation outcome of the fitness function, typically the combined effect of objective function and constraint violation incurred. For EA in general, there are three major paradigms—genetic algorithm[18], evolutionary programming[19], and evolution strategy[20]. The key genetic operators used in EA are crossover, mutation and selection. Briefly, crossover (or recombination) is a reproduction process of parents, mutation is to alter genetic information of parents, and selection is to choose offspring for survival, as shown in Figure 4.3. The major discrepancies among genetic algorithm, evolutionary programming and evolution strategy can be analyzed from their data representation, degree of importance of crossover and mutation, and the approach of selection. Owing to the multidimensional, nonlinear, mixed continuous-

Figure 4.3 Typical evolutionary loop of evolutionary algorithm

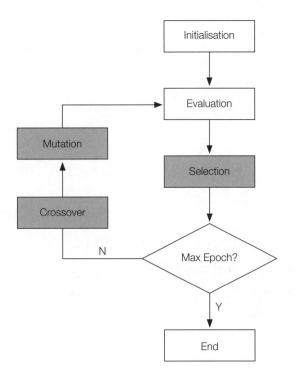

discrete and highly constrained nature of HVAC simulation models, EA is very effective to determine the optimal or near-optimal along the rugged search landscape.

3.2 Design of Piping Network for District Cooling System

As recognised, a major constraint in introducing district cooling is the high investment cost for the distribution network. In a district, how the piping network connecting the consumer buildings would affect the lengths of the pipe segments with different diameters as well as the pressure loss along the piping. The piping connection can be radial, tree-like circuit or a mix of them. Optimal configuration can result in both minimum piping initial cost and minimum pumping cost. Since there are an enormous number of different combinations of the piping configuration, it is infeasible to evaluate each individual case by exhaustive approach. In this example, genetic algorithm (GA) is applied to determine the optimal/near-optimal configuration of the piping network for a DCS[21].

A hypothetical site of reclaimed land with different building mix including offices, hotels, retails, schools, hospitals and mass transit railway station is proposed in this study[22]. All are considered as the potential DCS customers. The locations of the consumer buildings (nodes) are fixed. The piping network is modelled by a graph with undirected links. Either radial or tree-shaped network or a mix of both are allowed. The objective is to find the optimal/near-optimal piping network configuration of a district cooling system that minimises the infrastructure (piping) cost compatible with the minimum pumping energy cost as listed in Eq. (1) below.

$$\text{min} \quad Z = \sum_{i=1}^{n-1} \sum_{j=i+1}^{n} C_{ij} x_{ij} + \sum_{k=1}^{8,760} \textit{Hourly Pumping Cost}_k \quad (1)$$

where,

C_{ij} : the cost of (i, j) pipe segment = $L_{ij} \times CPUL_{ij}$

L_{ij} : pipe length of (i, j) pipe segment

$CPUL_{ij}$: cost per unit length of (i, j) pipe segment which depends on the pipe diameter

x_{ij} : $\in \{0, 1\}$ is the decision variable

One of the crucial factors for a successful implementation of GA is the representation of an underlying problem by a suitable scheme. In the present study, integer-string representation is used to encode the piping network. With this integer-string representation, every possible link is assigned with an integer. The number of integers should be equal to $n - 1$, where n is the number of nodes in the piping network. This kind of encoding is uniformly redundant, i.e. each phenotype is represented by the same number of genotypes. An example of a piping configuration of problem size 9 with integer-string representation is illustrated in Fig. 4.4 and Table 4.2. In Fig. 4.4, node 1 is a DCS central plant which supplies chilled water to the consumer buildings (nodes 2 – 9) through a piping network. This configuration is encoded by an integer-string {1, 5, 8, 9, 23, 27, 34, 35} as listed in the shaded cells of Table 4.2.

Beginning with an initialisation process, a population of size 20 is generated randomly. Twenty number of different piping configurations are formed. The corresponding cost values are calculated by Eq. (1). After that, rank scaling was used to scale the raw scores (total cost) of the objective function based on the rank of each individual, rather than its score. The rank of an individual is its position in the sorted scores. The rank of the fittest individual is 1, the next fittest is 2 and so on. Rank fitness scaling can remove the effect of the spread of the raw scores. Then selection function is applied to choose parents for the next generation based on their scaled values. In the present study, stochastic uniform selection function is adopted. It lays out a line in which each parent corresponds to a section of the line of length proportional to its expectation. The algorithm moves along the line in steps of equal size, one step for each parent. At each step, the algorithm allocates a parent from the section it lands on until the required number of parents are obtained for reproduction of children candidates.

Figure 4.4 An example network with integer-string {1, 5, 8, 9, 23, 27, 34, 35}

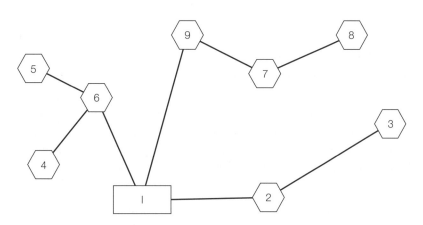

Table 4.2 Integer-string encoding

Integer label	Link	Integer label	Link	Integer label	Link
1	1-2	13	2-7	25	4-8
2	1-3	14	2-8	26	4-9
3	1-4	15	2-9	27	5-6
4	1-5	16	3-4	28	5-7
5	1-6	17	3-5	29	5-8
6	1-7	18	3-6	30	5-9
7	1-8	19	3-7	31	6-7
8	1-9	20	3-8	32	6-8
9	2-3	21	3-9	33	6-9
10	2-4	22	4-5	34	7-8
11	2-5	23	4-6	35	7-9
12	2-6	24	4-7	36	8-9

The new generation is produced by three operators: elitism, crossover and mutation. For elitism, it retains a certain number of individuals which are guaranteed to survive to the next generation. In this study, elite count is set to 1, i.e., the best solution in the current generation is carried forward to the next generation. Crossover fraction of 0.4 is used to specify the fraction of the next generation, other than elite individuals, that are produced by crossover. Scattered crossover operator is used to create a random binary vector which selects the genes where the vector is a 1 from the first parent, and the genes where the vector is a 0 from the second parent. Then a new child is formed by combining these genes. The remaining individuals are produced by mutation. Mutation function makes small random changes in the individuals of the population which provides genetic diversity and a broader search space for GA. Once the new populations are formed, the optimisation process repeats until termination criterion is met, e.g. a pre-set maximum number of generation of evaluation is exceeded.

The performance of GA on the optimisation of the piping network is investigated through a number of simulations. A district with problem sizes of 9 and 17 are used respectively. That is one DCS central plant serving eight consumer buildings in the first case and in the second case, one DCS central plant serves 16 consumer buildings. For each run, the numbers of generation are 200 and 1,000 for the cases with problem sizes of 9 and 17, respectively. The simulated results over 10 runs are listed in Table 4.3.

In Table 4.3, the best fitness value represents the minimum total cost of an optimal/ near-optimal piping configuration. For problem of size 9, it can be seen that the best one is $1,338 \times 10^6$ at simulation run no. 6. The first-hit generation number (i.e. the generation number at which the best solution first appears) is 150 and the layout of this best piping configuration is plotted in Figure 4.5. The result for the problem of size 17 is presented in a similar format. Figure 4.6 shows the layout of the piping configuration and the corresponding best fitness value is $3,290 \times 10^6$ obtained at simulation run no. 2. The first-hit generation number is 772. It is noted that the "worst" best fitness value is $5,077 \times 10^6$ at simulation run no. 5 and the average value is $4,208 \times 10^6$. This reflects a decrease in performance of GA for increased problem size.

The results show that GA can be applied for searching the optimal/near-optimal piping configuration in a DCS within acceptable computational timeframe. The methodology developed here can assist the engineers to design an optimal/near-optimal piping configuration of a DCS. Although, as mentioned above, the performance of searching the optimal/near-optimal configuration is reduced for increased problem size, there are potential areas for future work such as incorporating local search technique for improving the performance of GA for optimisation of piping network in DCS.

Table 4.3 Simulated results over 10 runs for problems of sizes 9 and 17

Run no.	Problem Size: 9		Problem Size: 17	
	Best Fitness Value (x10⁶)	First Hit Generation No.	Best Fitness Value (x10⁶)	First Hit Generation No.
1	1,381	187	3,577	935
2	1,447	157	3,290	772
3	1,367	188	3,592	896
4	1,367	57	3,656	970
5	1,591	185	5,077	976
6	1,338	150	4,453	907
7	1,339	185	4,790	614
8	1,601	192	4,203	998
9	1,435	105	4,748	914
10	1,339	153	4,698	929
Min.	1,338	57	3,290	614
Max.	1,601	192	5,077	998
Average	1,420	156	4,208	891

Figure 4.5 Optimal piping configuration for problem of size 9

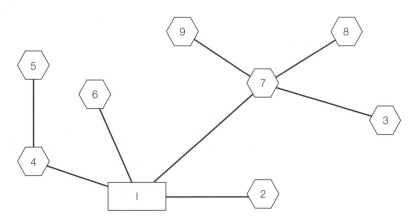

Figure 4.6 Optimal piping configuration for problem of size 17

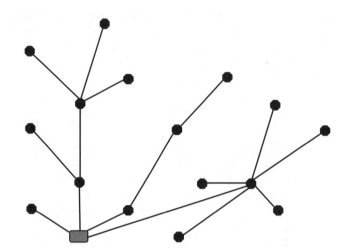

3.3 *Energy Management of Central HVAC System for Wider Coverage*

3.3.1   System description and EMO identification

The second case study is about a central HVAC system providing district chilled water supply for the corresponding air handling units (AHUs) in the five subway stations in Hong Kong[23]. The plant component simulation program TRNSYS[24] is used to model the entire chiller plant with total cooling capacity of 6,000 TR. The whole simulation plant model is divided into water side system and air side system, as shown in Figure 4.7. The water side system consists of six numbers of 1,000 TR water-cooled chillers, the associated chilled water pumps, the corresponding heat rejection equipment (including condenser water pumps, heat exchangers and sea water pumps), and the differential pressure bypass circuit. The air side system includes AHUs, the associated outdoor air and exhaust air fans, and free cooling fans. In order to respond to different cooling loads throughout a year, the plant model includes the control for automatic staging of chillers and associated pumps, and air side free cooling mode in case of suitable outdoor air enthalpy. The TRNSYS assembly panel of this central HVAC system is also shown in Figure 4.7.

In this application, the chilled water supply temperature set point of the central chiller plant is identified as the possible EMO for optimisation. This is EMO Cat I since no practical investment cost would be involved. In the existing plant operation,

Figure 4.7 Central HVAC system on assembly panel of TRNSYS

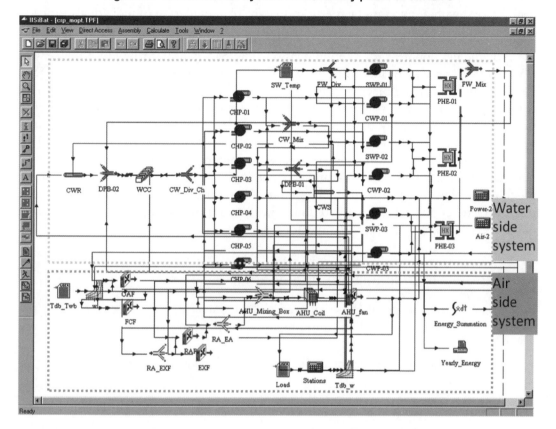

the chilled water supply temperature is constantly set at 7.2°C throughout the year, and no reference information is available for temperature reset purpose. In order to implement EMO Cat I and achieve more effective energy management, the total energy consumption of all the involved HVAC equipment should be minimised at an optimal chilled water supply temperature. The proposed reset scheme would be on monthly basis.

3.3.2   Simulation-optimisation development

This simulation-optimisation platform is developed by linking the plant simulation software TRNSYS with the EA written in MATLAB, both under the Windows environment of a conventional personal computer. For EA, the paradigm of evolutionary programming is applied, real-valued parameters are directly used, and no crossover process is involved. In the process of selection and mutation, the approach of elitism was adopted. The individual with the best fitness, i.e. the minimum monthly energy consumption, is selected and maintained for next epoch without any change, while the rest is varied with a mutation factor. In this study, the population of individuals for each epoch is ten, and the maximum number of epoch is 15.

For the development of plant simulation, TRNSYS is used to build up the central HVAC plant model, by including all the equipment components, and the related input parameters. The objective function is the monthly energy consumption, which is the key output of this simulation model. The fitness value is directly resulted from the number and efficiency of operating equipment to handle the hourly cooling load within that month. The energy consuming components include the chillers, chilled water pumps, condenser water pumps, sea water pumps, AHU fans and cooling coils, outdoor air fans, exhaust air fans, and free cooling fans. Hourly analysis is applied for the monthly simulation, where 744 hours (i.e. 31 days) for January, March, May, July, August, October and December; 720 hours (i.e. 30 days) for April, June, September and November; and 672 hours (i.e. 28 days) for February.

3.3.3   Results from simulation-optimisation

In Figure 4.8, the change of the optimised chilled water supply temperature with the epoch of EA is shown. Nearly all months, the chilled water supply temperatures are

Figure 4.8 Optimal chilled water supply temperature vs. epoch

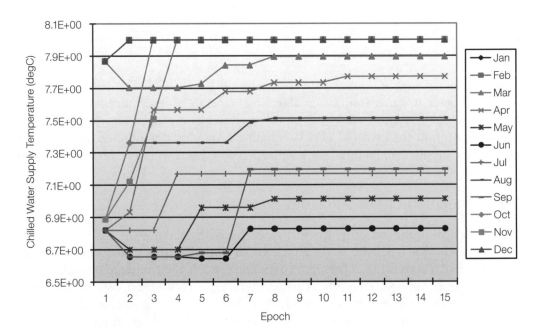

Figure 4.9 Minimum monthly energy consumption vs. epoch

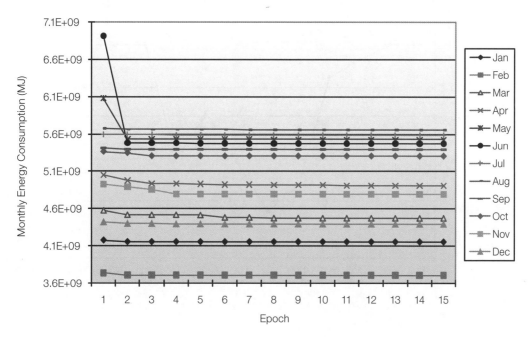

converged after the eighth epoch, this is also supported from the change of minimum monthly energy consumption against the epoch as shown in Figure 4.9. This shows that the proposed number of epoch is enough to provide the satisfactory and convergent results for this case.

The monthly optimal chilled water supply temperatures are summarised and illustrated in Figure 4.10. Compared with the existing constant set point 7.2°C throughout a year, the general trend is that higher chilled water supply temperatures can be set from October to April (i.e. the autumn, winter and spring in Hong Kong), while lower temperatures are required for May and June, as the humid weather will generate a comparatively high latent load from the outdoor air in Hong Kong. For the hottest months from July to September, the original set point 7.2°C is basically satisfactory in general.

Based on the optimal chilled water supply temperatures in different months, the corresponding monthly energy consumptions of the central chiller plant of the subway stations are summarised in Figure 4.11. The profile follows the climatic and environmental changes in Hong Kong, with the highest value in August (i.e. the summer) and the lowest value in February (i.e. the winter).

Figure 4.10 Monthly optimal chilled water supply temperature

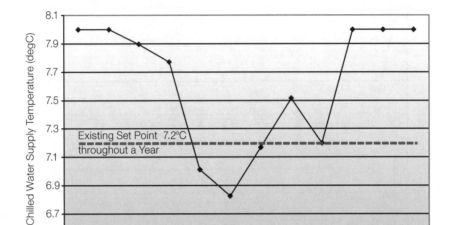

Figure 4.11 Monthly energy consumption of central HVAC system

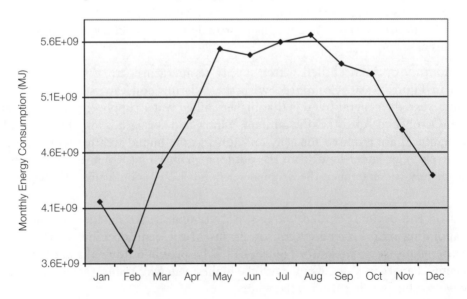

If there is building management system (BMS) for the HVAC installations, the developed information of the reset scheme can be used as a reference for the operating chilled water supply temperature in different months. The log data from BMS can then be used to review and verify the proposed optimal values. Through this research work, the monthly optimisation of chilled water supply temperature has been demonstrated. In effect, the analysis can be down to the weekly optimisation, and more comprehensive and frequent information can be provided for chilled water temperature reset, thus better energy management can be achieved.

The effectiveness of this simulation-optimisation approach has been demonstrated through the development of the monthly reset scheme of the chilled water supply temperature of a local central HVAC system. For further application, the supply air temperature is another possible parameter that can be optimised for better energy management. In fact the current performance-based optimisation method can be used to handle a variety of scenarios of the HVAC and building services systems in terms of energy conservation. In this case study, EA has demonstrated its capability in locating the feasible space more effectively, perceiving the searching topology more readily, and enhancing the searching rate more efficiently.

4 Conclusion

In this chapter, energy conservation measures and the common central HVAC development in Hong Kong are discussed. The two voluntary schemes, HKEERSB and HK-BEAM, demonstrate the commitment and participation of both the government and building sector in energy conservation in buildings. The former addresses a robust framework for buildings to meet the minimum energy requirements, while the latter provides a direction to pursue higher achievement in energy saving and sustainability. Besides new building projects, more and more existing buildings are willing to join in these registration schemes in order to benchmark the existing performance and seek for further improvement for sustainability purpose.

In view of building energy consumption, HVAC takes a major portion, almost a half. Suitable identification of energy conservation opportunities and design of central plant for HVAC systems is essential to achieve the goal of energy conservation. Centralised HVAC systems would undoubtedly inherit the advantages of better overall energy performance and efficiency throughout a year in the subtropical Hong Kong. Apart from the conventional approach of water side and air side design, district cooling system is possible to be applied in a large-scale HVAC design. With mass production in giant chillers and diversion of thermal loads, it would be an energy efficient scheme to provide chilled water to a number of buildings and a wide district.

In designing the distribution network or energy management of supply temperature for a large-scale HVAC system, effective simulation-optimisation method is crucial to handle different design scenarios and operation requirements. Owing to the nonlinear, multidimensional, multimodal and mixed continuous-discrete nature of the HVAC problems described by a complex of mathematical expressions or plant simulation model, evolutionary algorithm is found to be effective to determine the global optimal. Example cases about the design and energy management of central refrigeration systems are presented, showing the modeling techniques for HVAC problems and optimisation method by using evolutionary algorithm. The simulation-optimisation approach is promising to tackle the HVAC problems from a life-cycle perspective, and its application is not limited in buildings but beyond.

Notes

1. Electrical and Mechanical Services Department, The Government of the Hong Kong Special Administrative Region. *Promoting Energy Efficiency and Conservation.* www.emsd.gov.hk/emsd/eng/pee/index.shtml

2. Electrical and Mechanical Services Department, The Government of the Hong Kong Special Administrative Region. *HK Energy Efficiency Registration Scheme for Buildings: Publications.* www.emsd.gov.hk/emsd/eng/pee/eersb_pub_cp.shtml

3. Electrical and Mechanical Services Department, The Government of the Hong Kong Special Administrative Region. *Energy Management.* www.emsd.gov.hk/emsd/eng/pee/em.shtml

4. Electrical and Mechanical Services Department, The Government of the Hong Kong Special Administrative Region. *Energy Consumption Indicators and Benchmarks.* www.emsd.gov.hk/emsd/eng/pee/ecib.shtml#

5. HK-BEAM Society. www.hk-beam.org.hk/general/home.php

6. HK-BEAM Society. (2005). *Enhancing Hong Kong's Built Environment.*

7. Rishel, J. B. (1983). Energy conservation in hot and chilled water systems. *ASHRAE Transactions, Part II B*, 352–367.

8. Lam, C. (2004). The Development of District Cooling Systems in Hong Kong, *Proceedings of the 2004 Shenyang-Hong Kong Joint Symposium on Healthy Building* in *Urban Environment* (pp. A48–A56), 29–30 July 2004, Shenyang, China.

9. Vadrot, A., Delbès, J. (1999). *District Cooling Handbook.* European Marketing Group District Heating and Cooling.

10. Hooke, R., & Jeeves, T. A. (1960). Direct search solution of numerical and statistical problems. *Journal of Association of Computing Machinery, 8*, 212–229.

11. Nelder, J. A., & Mead, R. (1965). A simplex method for function minimisation. *The Computer Journal, 7*, 308–313.

12. Wright, J. A. (1986). *The optimised design of HVAC systems*. PhD thesis, Loughborough University of Technology. UK; Wright, J. A., & Hanby, V.I. (1987). The formulation, characteristics, and solution of HVAC system optimized design problems. *ASHRAE Transaction, 93*(2), 2133–2145.

13. Kirkpatrick, Jr. S., Gelatt, C., & Vecchi, M. (1983). Optimization by simulated annealing. *Science, 220*(4598), 498–516.

14. Glover, F. (1989). Tabu search—Part I. *ORSA Journal on Computing, 1*(3), 190–206; Glover, F. (1990). Tabu search—Part II. *ORSA Journal on Computing, 2*(1), 4–32.

15. Dorigo, M., & Maria, G. (1997). Ant colony system: a cooperative learning approach to the traveling salesman problem. *IEEE Transaction Evolutionary Computation, 1*, 53–66.

16. Kennedy, J., & Eberhart, R. C. (1995). Particle swarm optimization. In *IEEE International Conference on Neural Networks, Vol. IV.* (pp. 1942–1948). Perth, Australia, November 1995.

17. Koeppel, E. A., Mitchell, J. W., Klein, S. A., & Flake, B. A. (1995). Optimal supervisory control of an absorption chiller system. *HVAC &R Research, 1*(4), 325–342.

18. Holland, J. H. (1975). *Adaptation in natural and artificial systems*. University of Michigan Press, Ann Harbor; Goldberg, D. E. (1989). *Genetic algorithms in search, optimization and machine learning*. Addison Wesley Publishing Company.

19. Fogel, L. J. (1964). *On the organization of intellect*. PhD Thesis, University of California at Los Angeles; Fogel, D. B. (1992). *Evolving artificial intelligence*. PhD Thesis, University of California, San Diego.

20. Rechenberg, I. (1973). *Evolutionsstrategie: Optimierung technischer systeme nach prinzipien der biologischen evolution*. Frommann-Holzboog, Verlag, Stuttgart; Schewfel, H.-P. (1981). *Numerical optimization of computer models*. Wiley, Great Britain.

21. Chan, A. L. S, Hanby, V. I., & Chow, T. T. (2007). Optimization of distribution piping network in district cooling system using genetic algorithm with local search. *Energy Conversion and Management, 48*(10), 2622–2629.

22. Chow, T. T., Au, W. H., Yau, R., Cheng, V., Chan, A. L. S., & Fong, K. F. (2004). Applying district-cooling technology in Hong Kong. *Applied Energy, 79*(3), 275–289.

23. Fong, K. F., Hanby, V. I., & Chow, T. T. (2006). HVAC system optimization for energy management by evolutionary programming. *Energy and Buildings, 38*(3), 220–231.

24. TRNSYS A Transient System Simulation Program. (March 2000). *Reference Manual, Volume I, Chapter 2*. Solar Energy Laboratory, University of Wisconsin-Madison, Madison.

5

Fire Services Engineering and Performance-based Design

Fire services installations are highly important in modern buildings, especially for special and mega buildings having difficulties in conforming with current prescriptive code, standard and legislative requirements in the Hong Kong Special Administrative Region (HKSAR). Inadequate fire prevention and protection, or misuse of general design criteria may lead to an uncontrollable situation and result in serious losses of properties and lives.

This chapter will refresh us on the current approval procedure, standard and legislative requirements, and the design approach of fire services installations in Hong Kong. A brief review of the fire services and safety and design concepts for modern buildings will also be presented.

Alan FONG, Zhang LIN and Tin Tai CHOW

Building Energy and Environmental Technology Research Unit
Division of Building Science and Technology
College of Science and Engineering
City University of Hong Kong

1 Design Approach of Fire Services Installation in Hong Kong

Codes of Practice for Minimum Fire Service Installations and Equipment and Inspection, Testing and Maintenance and Equipment (COP FSI) in HKSAR[1] are basically prescriptive in nature, following those developed decades ago in the United Kingdom with some slight modifications under the Circular Letters issued by the Fire Services Department. Approval of fire safety designs and inspection of the buildings upon completion are held responsible by two government departments: BD and FSD. Current fire services building design should be submitted to the Buildings Department (BD) for approval. The requirements and installations of fire protection systems are monitored by the Fire Services Department (FSD). A pictorial presentation of the application procedure is shown in Figure 5.1 for general application in accordance with FSD Circular letter no. 1/2005.

The procedures for submission of FSI/314 and FSI/501 as stipulated in FSD circular letter no. 1/2005 was superseded by the 4/2008 with immediate effect. For facilitating a smooth transition for trade, the procedures were repealed on 31 October 2008. The major change was to introduce a new "Help Desk" services in lieu of the stage 1—Design stage for FSI plan submissions as stipulated in FSD circular letter no. 1/2005.

For those buildings having difficulties in compliance with the prescriptive fire safety codes, the fire engineering approach has been accepted by BD as an alternative route since 1998. The Practice Note for Authorised Persons and Registered Structural Engineers (PNAP) 204[2] provides guidance on fire engineering approach (FEA) for the design of new buildings or alteration and addition works in existing buildings. The approval procedure of FEA is shown in Figure 5.2, which is extruded from the Appendix of this PNAP.

All regulations and ordinance in this chapter are quoted up to August 2008.

Figure 5.1 Approval procedure for fire service design under FSD circular letter no. 1/2005

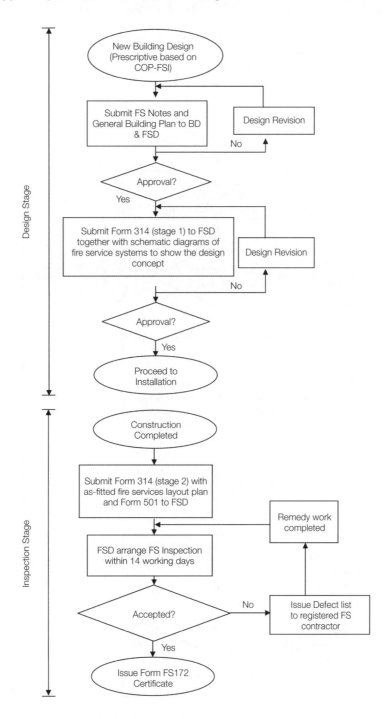

Figure 5.2 Overview of fire engineering approach design
(extruded from the Appendix of PNAP 204 "Guideline to Fire Engineering Approach")

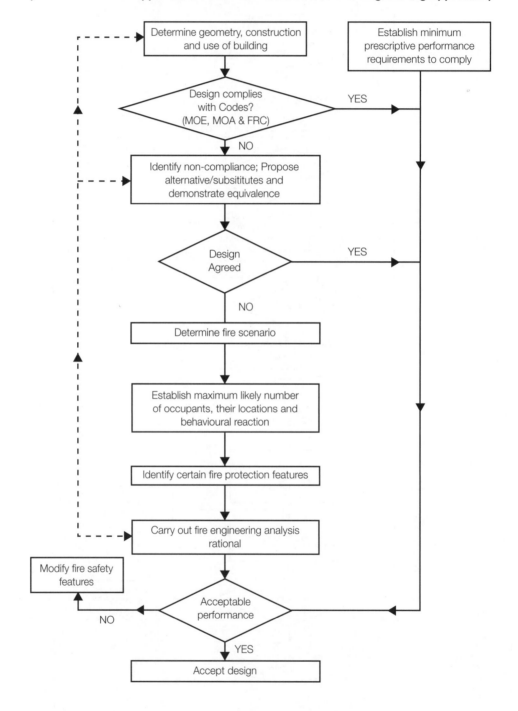

2 The Current Hong Kong Buildings Ordinance, Regulations and Codes of Practice Related to Fire Safety Matters

The objective of building regulations is to protect life and prevent injury in the event of fire, and to protect goods, properties and building content. Codes of practice have been developed to provide the technical basis for such regulations.

Under the current Hong Kong building legislation system, the Authorised Person (AP) is to ensure the design and construction of building works comply with Buildings Ordinance and Regulations, and the building services consulting engineer (BSE) works as an assistant or a representative of the authorised person in certain governmental procedures and application.

An authorised person can be a registered architect (AP-List I), a registered structural engineer (AP-List II), or a registered building surveyor (AP-List III). An "authorised person" is legally defined in the Hong Kong Buildings Ordinance Cap. 123.

BSE could also be Registered Professional Engineers (RPE) under Engineers Registration Ordinance in 1990 for certain governmental submission. For example, BD asks for RPE to sign the Overall Thermal Transfer Value (OTTV) calculation for General Building Plan submission as stated on PNAP 172.

2.1 The Present Hong Kong Buildings Ordinance and Regulations

According to Buildings Ordinance, Cap 123, the present Hong Kong Building Regulations are mainly contained in various sections of Chapter 123 of the Laws of Hong Kong. Those sections related to fire safety matters are shown in Table 5.1.

2.2 Other Hong Kong Regulations Relating to Fire Services

Other important regulations relating to fire services in buildings include:

1. Chapter 95—Fire Services Ordinance
    1.1 Chapter 95A—Fire Services (Installation Contractors) Regulations
    1.2 Chapter 95B—Fire Services (Installation and Equipment) Regulations

1.3 Chapter 95C—Fire Services Department (Reports and Certificates Regulations)

1.4 Chapter 95D—Fire Services Department (Welfare Fund) Regulations

2. Chapter 172—Places of Public Entertainment Ordinance

3. Chapter 295—Dangerous Goods Ordinance and Regulations

4. Chapter 502—Fire Safety (Commercial Premises) Ordinance

5. Chapter 572—Fire Safety (Buildings) Ordinance

Table 5.1 Hong Kong Building Regulations in Chapter 123 related to fire safety matters

	Hong Kong Building Regulations, CAP 123	Areas of Relevance to BSE (FS)
1	CAP 123A Building (Administration) Regulations	BSE (FS) to supply AP with FS Notes for inclusion in architectural plans to BD/FSD.
2	CAP 123 B Building (Construction) Regulations	BSE (FS) to be familiar with Part XVI—Fire resisting construction of Cap. 123B, and BS 476: Part 8 for fire resistance of building materials. BS476 Parts 20–24, and HK COP for Fire Resisting Construction.
3	CAP 123 D Building (Escalators) Regulations	Relevant to Registered Escalator Engineer and BSE for design, installation and construction.
4	CAP 123 E Building (Lifts) Regulations	Relevant to Registered lift engineer and BSE for design, instillation and construction.
5	CAP 123 F Building (Planning) Regulations Pursuant to Regulation 41(1) of these regulations, the Code of Practice on Provision of Means of Escape in Case of Fire and Allied Requirement to be complied with.	BSE to define in FS Notes the form of lighting (e.g. supported by emergency power generator) and ventilation (e.g. mechanical ventilation for staircases/office occupancy.) BSE (FS) should have a general concept on the Code of Practice of Means of Escape.
6	CAP 123 J Building (Ventilation Systems) Regulations	Supplemented by ventilated circular letters issued by FSD, relevant to BSE (HVAC).
7	CAP 123 K Building (Oil Storage Installations) Regulations	BSE for FS requirements

The above regulations are supplemented by circular letters. Practice Notes for Authorised Persons and Registered Structural Engineers are issued occasionally by the Building Authority.

Pursuant to Chapter 95, FSD has released the following to supplement and amplify the FS requirements:

1. Codes of Practice on Minimum Fire Services Installations and Equipment, and Inspection and Testing of Installation and Equipment, the latest versions were released in July 2005 abbreviated as COP-FSI and COP-I&T. These codes are pursuant to Section 16(1)(b)(ii) and 21(6)(d) of Buildings Ordinance, Cap 123 and Regulation 10 of Fire Service (Installations and Equipment) Regulations of Fire Services Ordinance, Cap. 95.

2. Occasional FSD circular letters issued to the local fire professional.

3. F. S. ventilation circular letters issued to the local fire professional.

4. Regulation sheets related to the requirements for places of public entertainment (e.g. restaurants), requirements for installations related to dangerous goods (the Dangerous Goods Technical Sheets—DG/TS series sheets) are given by FSD for projects from case to case.

5. Standing orders and circulars which are issued internally within FSD are not disclosed to the public. Many of such issues are related to building plan approval and FS installation requirements.

According to Clause 1.3 of the Codes of Practice, it is stipulated that the Director of Fire Services (or his delegated representative officer) may vary any requirement of the code in specific cases. This gives FSD the final power to specify FS provisions to individual buildings.

In addition to the above issues, FSD is also enforcing the following Rules, Codes, and Standards for commonly used systems listed in Table 5.2, as referred to the COP FSI [1].

FSD may accept an alternative of equipment and installation as well as testing standards recognised and certified by the laboratories/organisations listed in FSD Circular Letters on Fire Test Facility. The use of alternative equipment/standard shall be stated in the FS Notes in drawings submitted to the Building Ordinance Office (BOO)/ FSD.

Furthermore, FSD issues a series of Fire Protection Notice and educational leaflets and pamphlets. Dangerous Goods (DG) General as stipulated on Fire Protection Notice No. 4 is of particular concern as it lists the requirements for DG licensing, penalty clauses, and exempted quantities for common dangerous goods.

Table 5.2 Rules/Codes/Standards related to commonly used fire services systems

FS Installation	Code/Rule/Standard
1. Sprinkler system	Fire Offices' Committee (FOC) Loss Prevention Council (LPC) Rules for Automatic Sprinkler Installations incorporating BSEN 12845 and Technical Bulletins, as amended by FSD Circular Letter No. 3/2006 (formerly the BS 5306: Part 2 Fire extinguishing installations and equipment on premises: Part 2) For building plans with first Building Ordinance Office (BOO) submission made on and after 1 January 2007, the LPC Sprinklers Rules, comprising BSEN 12845 with local modification. Before that day, the submission shall be based on BS 5306: Part 2 with local modification. (Note: FSD may accept deviation from LPC Sprinkler Rules for Improvised Sprinkler Installations, which are installed for enhancing safety of old commercial buildings based on individual cases)
2. Automatic fire alarm and detection system	Rules for Automatic Fire Alarm Installations for the Protection of Property for property protection, which calls for BS5839: Part 1 with local modification under FSD Circular letter no. 1/2002.
3. Smoke Control System	British Standard 5588 part 4—Code of Practice for smoke control using pressure differentials incorporating amendment no.1 and corrigendum no.1 1998 edition as amended by FSD Circular letter no. 2/2006.
4. Roller Shutters / Fire Doors	BD now specifies fire shutters to comply with the relevant parts of BS476. Notes I. BD/FSD accepts the standards and test certification of some reputable testing laboratories of U.K., Australia, New Zealand, etc. as listed in the latest issue of FSD Circular Letter on Fire Test Facility. II. FSD also accepts National Fire Code (NFC) 80 and National Fire Protection Association (NFPA) for fire doors and windows, covering shutters as well. All fire shutters proprietarily made to be submitted separately to FSD and BD for acceptance for each project.
5. External Drenchers	Separate submission to FSD on individual case basis.
6. Carbon Dioxide System	NFPA 12
7. Clean agent systems	NFPA 2001
8. Water spray fixed system for fire protection	NFPA 15

2.3 Other Codes of Practice Published by Buildings Department Relating to Fire Safety Design

1. Code of Practice for the Provision of Means of Access for Fire Fighting and Rescue Purposes (MOA) published by Building Authority[3]—this Code provides guidance to assist in firefighting and in life saving by ensuring adequate access for firefighting personnel in the event of fire and other emergencies as well.

2. Code of Practice for Fire Resisting Construction (FRC) issued by BD[4]—this Code is to announce provisions for the protection of buildings from the effects of fire by inhibiting the spread of fire and ensuring the integrity of the structural elements of building. For example, suitable form of fire stop or fire dampers should be provided for E&M services passing through opening at a compartment wall or floor.

3. Provision of Means of Escape in Case of Fire (MoE) published by Building Authority [5]—this Code is to announce provisions for the protection of buildings from the effects of fire by providing adequate means of escape in the event of fire and other emergencies as well.

2.4 Waterworks Ordinance

The Chapter 102 Waterworks Ordinance and the Hong Kong Waterworks Standard Requirements: Sheet No. 9 for installation of a fresh/salt water fire services are applicable to Fire Services Installation.

Water Supplies Department (WSD) occasionally issued circular letters to licensed plumbers. Letters of particular concern are on the approval of water supply fittings, and on Point Penalty System for Defects in Plumbing Works carried out by licensed plumbers. WSD published "A guide to the Preparation of Plumbing Proposals" (for submission to Water Authority for approval) in December 2006. This guide applies to water-based Fire Services Installation (FSI).

2.5 Miscellaneous Regulations and Other Bodies Having Jurisdiction

1. Civil Aviation Authority (e.g. for fire services installation for helipads)

2. Electricity supply ordinance (Chapter 103 of Laws of Hong Kong) and the power supply companies—CLP Power Hong Kong Ltd. and Hong Kong Electric Co. Ltd.

3. Environmental Protection Department (e.g. for emission of flue gas from diesel pumps and diesel emergency generators of FS Installations)

4. Education Department (e.g. for fire fighting equipment for school laboratories)

5. Hong Kong Telecom (for direct telephone line connection to Fire Services Communication Centre, usually applied through the FS Contractor and Chubb carrying out maintenance for the control panel at the FS control centre).

6. Labour Department (HK Law Cap 59)—Factories and Industrial Undertakings Ordinance/Regulations (e.g. Fire Safety for industrial plants in operation)

7. Port Works (e.g. for fire safety of seaside pump house, requiring fire extinguishers, fire detectors sometimes)

8. Regional Services Department (and Licensing Authority) (e.g. for Hotel licensing FS requirements.) and Urban Services Department (e.g. for licensing requirements for related business).

9. The Hong Kong Radiation Board under the Medical and Health Department controlling ionisation smoke detector installation and storage.

10. The Fire Insurance Association of Hong Kong—This is the local channel to UK LPC adviser to its member insurance companies of FS requirements for buildings.

3 Fire Safety and Services Installation in Modern Building

As mentioned in Section 2, there are many Hong Kong Buildings Ordinance, Regulations and Codes of practice needed to be considered for fire safety. The following major facilities will be discussed related to fire safety and services installation.

3.1 Fire Services Installation

The following items required fire fighting and protection systems, and installations are specified in the FSD Codes of Practice (COP), known as the Code FSI.

    i. Automatic actuating devices

    ii. Automatic fixed installation other than water

    iii. Automatic fixed installation using water

    iv. Deluge system

    v. Drencher system

    vi. Dust detection system

    vii. Dynamic smoke extraction system

    viii. Emergency generator

    ix. Emergency lighting

    x. Exit sign

    xi. Fire alarm system

    xii. Fire control centre

    xiii. Fire detection system

    xiv. Fire hydrant/hose reel system

    xv. Fireman's lift

    xvi. Firefighting and rescue stairway

    xvii. Fixed automatically operated approved appliance

    xviii. Fixed foam system

    xix. Gas detection system

    xx. Gas extraction system

    xxi. Portable hand-operated approved appliance

    xxii. Pressurisation of staircase

    xxiii. Ring main system with fixed pump(s)

    xxiv. Sprinkler system

    xxv. Static smoke extraction system

    xxvi. Street fire hydrant system

    xxvii. Supply tank

    xxviii. Ventilation/air conditioning control system

    xxix. Water mist system

    xxx. Water spray system

    xxxi. Water supply

Various types of buildings should comply with the above specified systems and installations covered by the specification of FSI code (see Note 1).

3.2 The Concern of Means of Escape (MoE) in Modern Buildings

The guidance of assessing the requirements of MoE or the capacity or populations of various portions of a building, or the number of persons and population density within a building are mentioned in the local MoE code [5].

In order to provide means of evacuation for occupants in buildings in the case of fire, two design approaches of evacuation are available. These include traditional prescriptive approach as stipulated in the local code of practice and fire engineering approach. When the fire engineering approach is adopted, there is a large range of design parameters and systems to be considered, such as the various fire safety systems adopted to achieve the required safe egress time. This will be discussed in Section 4.

3.3 The Concern of Smoke Extraction System in Modern Buildings

According to the FSI Code, dynamic or static smoke extraction system should be provided for any fire compartment that exceeds the specified volume capacity (such as exceeding 7,000 cubic meters in hotel building and basement floor or exceeding 28,000 cubic meters in atrium floor) in the various types of modern building, where the aggregate area of openable window of the compartment does not exceed 6.25% of the floor area of that compartment, and the design fire load is likely to exceed 1,136 MJ/m^2. There are again two design approaches in the smoke extraction system design: the prescriptive method and the fire engineering approach.

The traditional prescriptive method results in a smoke extraction provision with the minimum extraction rate not less than eight air charges per hour of the total compartment volume.

The fire engineering approach estimates the required extraction rate based on the convective component of the design fire heat release rate, the type of smoke plume and the design smoke layer interface height at the concern protected area. The objective is to extend the safe egress time by maintaining a certain height of smoke clear layer or delaying the descent of smoke layer during evacuation. This approach should also take into consideration the buildings characteristics and its normal mode of function.

4 Performance-based Design Approach for Fire Services

Traditionally, the buildings ordinance and codes are prescriptive in nature. However, such codes cannot cover emerging and innovative technologies. Thus, prescriptive regulations may impose constraints which are not always appropriate to specific building projects.

In order to free designers from such constraints, to encourage innovation, and to facilitate trade, building codes accept the performance-based alternative.

For any building having difficulties in compliance with the prescriptive fire safety codes, the general procedure for Fire Engineering Approach (FEA) as shown in the Flow Chart of Figure 5.2 has been accepted by BD since 1998 as one possible methodology to process the performance-based design (PBD).

4.1 Implementation of Fire Engineering Approach in Hong Kong

A Fire Safety Committee (FSC) chaired by an Assistant Director was set up by BD in 1998 to consider fire safety designs with FEA. Other members included the Chief Building Surveyor responsible for fire, the other Chief Building Surveyors, the Chief Structural Engineers, a representative officer from FSD, and two experts in fire engineering who were not government officials.

There are so far no standard or structured methods for assessing these alternative designs. Those PBD approaches used overseas are currently applying in Hong Kong. In most of the submitted reports, the parts deviated from the prescriptive codes will be assessed. Non-compliance with code FRC (See Note 4) for glazing walls in a double-skin façade is an example. In fact, the FEA applications in most projects are to demonstrate their equivalence to the prescriptive codes.

4.2 Performance-based Design for Mean of Escape Route

The design approach for escape routes of a Hong Kong building can be based on both the prescriptive approach and the fire engineering approach. In prescriptive approach, the designers shall follow the guidelines in the codes of practice (see Note 3), practice notes for authorised persons, circular letters, etc. For most buildings in Hong Kong, the escape routes design are still based on these codes of practice to obtain the "deem to satisfy" provision. One of the code of practice (see Note 1) issued by FSD and the following three codes of practice issued by BD shall be in full compliance with in designing escape route.

Prescriptive code	Briefing of the code requirement
FRC code [4]	The provisions for the protection of building and escape route using suitable non-combustible materials (which possess a specified fire resistance period for different construction elements and resisting the action of fire) are described. It also stipulates the integrity, stability and insulation requirements for the building elements.
MoE code [5]	Prescribed figures are provided for the building designer to determine the occupant density of the building, number of staircases in both sprinklered and non-sprinklered buildings, discharge values, travel distances, and staircase width, etc.
CIBSE Guide E [6]	The evacuation time for each storey to a protected area (e.g. the staircase leading to the exit) should be within a notional period of 2.5 minutes (150 s) for non-sprinklered buildings.

Under the prescriptive approach, the escape route design is based on the requirements stated in the code of practice without questioning the actual performance of these escape routes and its interaction with the occupants and other building features. A design following these codes of practice is presumed to provide sufficient protection to the occupants and the building in the case of fire.

However, for buildings with special features, FEA has to be adopted to design the escape route. A common practice is to use the timeline approach to estimate the evacuation time of the building occupants.

4.3 Performance-based Design Using of the Timeline Approach

In the timeline approach, fire engineers will study the probable fire scenarios by considering the specific features/functions of the building, and the characteristics of the occupancy. The appropriate active fire service installation will be recommended. The fire engineers will also determine the maximum design population and the proposed means of escape, together with the fire location and the worst credible fire scenarios. A design fire size will be determined based on the chosen fire location. These are often subjects leading to strong arguments and the NFPA, CIBSE and journal papers will be the reference materials in support of the rationale. Wind and stack effects affecting the fire development and smoke movement will also be considered. All factors concerned will be used as input parameters for the fire and smoke spread calculation.

The calculation may be based on empirical equations provided in the standards and design guides such as CIBSE TM19 or CIBSE Guide E. An alternative is to use zone models such as CFAST[7] or field models such as Fire Dynamics Simulator (FDS)[8] to estimate the tenability conditions of the building in case of fire. The untenable conditions are in term of smoke layer interface height and thermal radiation from fire and enclosure temperature. The time from ignition to the occurrence of untenable conditions will be estimated and taken to be the available safe egress time (ASET) for the building occupants.

In estimating the evacuation time of the occupants, designers or engineers will first determine the ultimate place of temporary safety. Empirical equations from the Handbook of Society of Fire Protection Engineers (SFPE) or evacuation software such as SIMULEX, STEPS and EXODUS will be used to calculate the total evacuation time/travelling time of the occupants. After the congested areas, dead-ends and extended travelling time of the occupants are identified, the difference of ASET and the required safe egress/escape time (RSET) of the occupants will then be calculated. If ASET is larger than RSET, the escape route design will be considered as appropriate. One of the comprehensive documents available for providing a systematic approach to the calculation of escape time is the British Standard BS7974:2002[9].

The following formula is proposed in BS7974:2002:

$$RSET = \Delta t_{det} + \Delta t_{alarm} + (\Delta t_{pre} + \Delta t_{travel})$$

where Δt_{det} is the time from ignition to detection determined by engineering tool such as the empirical model given by Alpert, Δt_{alarm} is the time from detection to a general alarm, Δt_{pre} is the pre-movement time for the building occupants, and Δt_{tra} is the travel time of the building occupants which can be determined by the evacuation software such as SIMULEX, STEPS[10].

5 Examples for Applying Computational Fluid Dynamics in Studying Flame and Smoke Propagation

In the following two examples, the large eddy simulation technique was applied to predict the spread of smoke and fire in the confined spaces: (1) public underground car park[11], and (2) public transport interchange (above ground)[12]. The simulation platform was the fire dynamics simulator, which was developed by the US National Institute of Standards and Technology (NIST).

Fire Dynamics Simulator (FDS) is a computational fluid dynamics (CFD) model of fire-driven fluid flow. The software solves Navier-Stokes equations appropriate for low-speed, thermally-driven flow numerically, with an emphasis on smoke and heat transport from fires. The incompressible Navier Stokes equations written in tensorial notations are:

$$\frac{\partial u_i}{\partial x_i} = 0 \qquad\qquad [1]$$

and

$$\frac{\partial u_i}{\partial t} + \frac{\partial u_i u_j}{\partial x_j} = \frac{1}{\rho}\frac{\partial p}{\partial x_i} + v\frac{\partial^2 u_i}{\partial x_j \partial x_j} \qquad\qquad [2]$$

where i and j stand for 1, 2 or 3 corresponding to the rectangular co-ordinates x, y and z; p is the static pressure; ρ is the density; u is the velocity and v is the kinematic viscosity of the fluid. The other equations of Navier-Stokes can be found on the website address: www.cfd-online.com/Wiki/ANavier-stokes_equations.

5.1 Public Underground Car Park

Figure 5.3 shows the physical layout of the car park, of dimensions 60m x 30m x 5m. The ventilation supply inlets and exhaust outlets are uniformly distributed at the ceiling level, as shown in Figure 5.4. There are two 10m wide vehicle exits at the opposite walls, together with a number of emergency exits for the occupants.

For the assessment of the mixing ventilation design scheme, the sources of ignition energy in vehicles were taken the same as those associated with structure fires. In this study, the modelled fire was initiated at the top of a vehicle. The fire was increased linearly up to 5MW within 100 seconds. Two case studies were performed using different ventilating rates at ambient temperature, i.e. 7 ACH for Case 1 and 14 ACH for Case 2. The supply velocities were 3m/s and 6m/s for Cases 1 and 2 respectively.

A comparison of the simulation results between Case 1 and Case 2 shows that the smoke and fire spread is significantly affected by the ventilation rate. A higher ventilation

Figure 5.3 3-D view of the physical model of the underground car park

Figure 5.4 Positions of supply inlets and exhaust outlets at ceiling Level

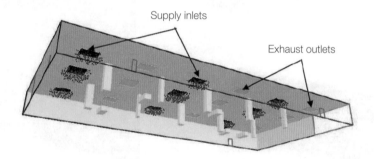

rate will intensify the fire and smoke propagation, as illustrated in Figure 5.5. In both cases, a mushroom cloud rises up to the ceiling and then spreads horizontally to the rest of the car park. The time for the smoke to reach the far end is about 20 seconds. Such a speedy spread gives alarm to the disastrous nature of these underground spaces.

Figure 5.6 shows the instantaneous temperature profiles respectively at 10, 20 40 and 100 seconds at 1.5m above the floor level. The average temperatures for Case 2 are clearly higher than for Case 1 at all time instants. Generally speaking, the temperature can become unbearable after 30s. The maximum fire temperature occurs at around t = 40s, reaching 965°C for Case 1 and 1,000°C for Case 2. The temperature then drops gradually with the decrease in combustion efficiency, caused by the buildup of smoke and the consequential reduction in oxygen supply. Comparing the temperatures at t = 40s and at t = 100s, it is found that for both cases, the regions affected by high temperature decreases as the time elapses. The doubling of the ventilation rate increases the overall temperature in the occupied zone significantly.

Figure 5.5 A comparision of smoke and flame propagation with time after the fire outbreak in underground car park for cases 1 and 2

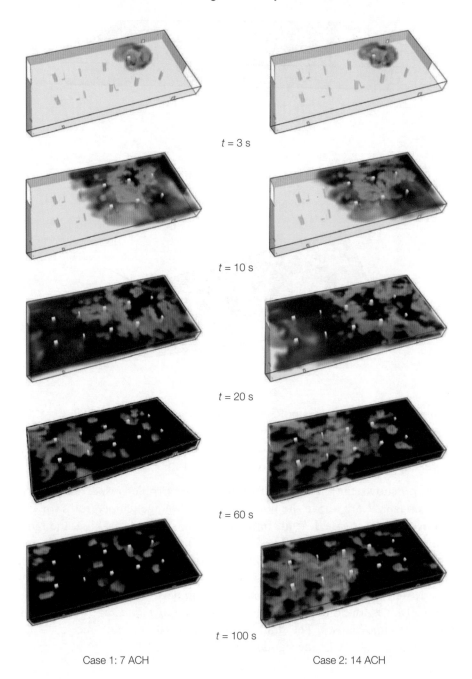

$t = 3$ s

$t = 10$ s

$t = 20$ s

$t = 60$ s

$t = 100$ s

Case 1: 7 ACH   Case 2: 14 ACH

Figure 5.6 Plan view of temperature distribution at 1.5 m above floor level (°C) in underground carpark

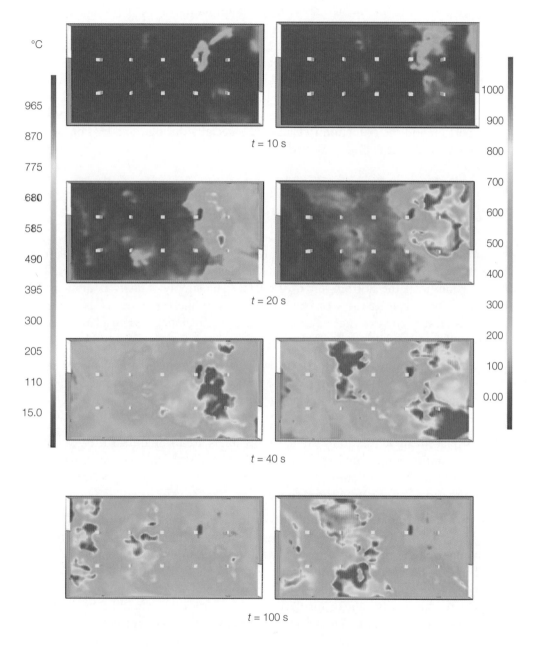

t = 10 s

t = 20 s

t = 40 s

t = 100 s

Case 1: 7 ACH   Case 2: 14 ACH

The above results show that in such a confined space the smoke and the flame can spread quickly and may fill up almost the entire space in less than one minute. This time span may be too short for the actuation of the automatic sprinkler protection system, and inadequate for the evacuation of the occupants. In this sense, the thread is high. For this specific floor plan layout, it will be desirable to increase the number of exits, and to shorten the vehicle exit routes. On the other hand, as more ventilating air tends to accelerate the fire spread and leads to a higher temperature environment, it becomes questionable whether the use of a higher ventilating rate for better indoor air quality is a good design solution. As a matter of fact, an increase of ventilation tends to worsen the smoke dispersion and the reduction in visibility, especially during the initial phase of the fire. The risk of flashover is also increased.

5.2 Public Transport Interchange (above ground)

This single floor Public Transport Interchange (PTI) (see Note 12) consists of several parked vehicles around a temporary internal shelter located at the centre of the PTI. An additional room is at the bottom left of the PTI. The supply velocity is 3m/s at room temperature. The exit is at the west side of the wall while the entrance is at the east. Various exits exist at the external boundary (building envelope). Exhausts are present at the ceiling and supply inlets are located at difference levels.

This study examines the effects of two different ventilation systems: mixing ventilation (Case A) and displacement ventilation (Case B), and their impacts on PTI fire safety. In a mixing ventilation system, supply and exhaust are both located at the ceiling (see Figure 5.7).

In a displacement ventilation system, supply inlets are located near the ground level while exhaust outlets are located at the ceiling (Figure 5.8). Displacement ventilation, as compared with mixing ventilation, is generally regarded as providing better indoor air quality and thermal comfort to a room space.

The profile of smoke propagation, according to the simulation results, is shown in Figure 5.9 and the temperature profile is shown in Figure 5.10. It can be observed that, by comparing the results of Case A and Case B, the smoke and fire are significantly increased in Displacement Ventilation during the first 20 seconds. This indicates that displacement ventilation will be abetting a fire spread. The finding unfortunately does not support the known benefit of displacement ventilation in providing better thermal comfort and indoor air quality. The selection between mixing or displacement ventilation thus has to consider not only health and comfort, but also life safety—how the provision will affect the growth and intensity of fire.

Figure 5.7 3-D view of the PTI for mixing ventilation system

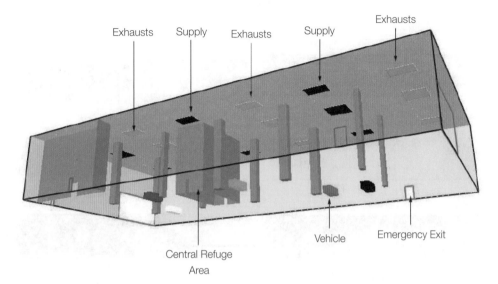

Figure 5.8 3-D view of the PTI for displacement ventilation (Supply Inlet at low level)

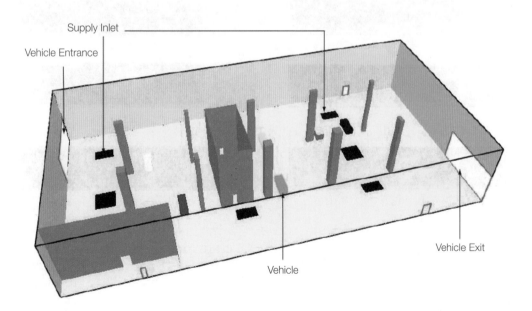

**Figure 5.9 Elevation view of smoke and fire dispersion in Case A and Case B
for aboveground PTI**

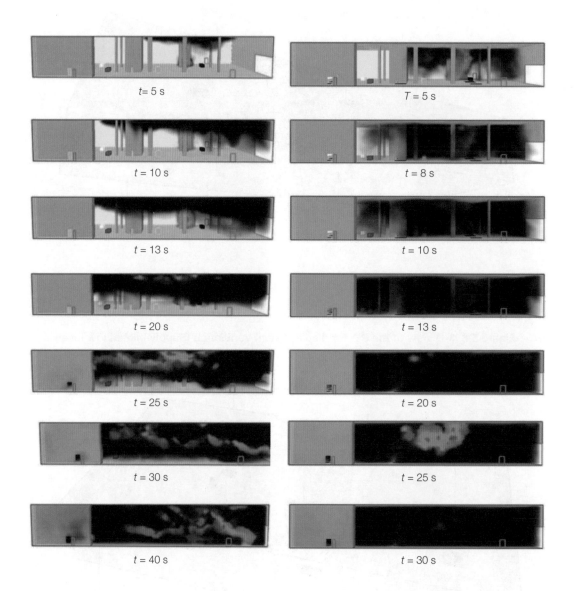

Case A: Mixing Ventilation System   Case B : Displacement Ventilation System

Figure 5.10 Temperature distribution of the aboveground PTI in Case A & Case B for PTI

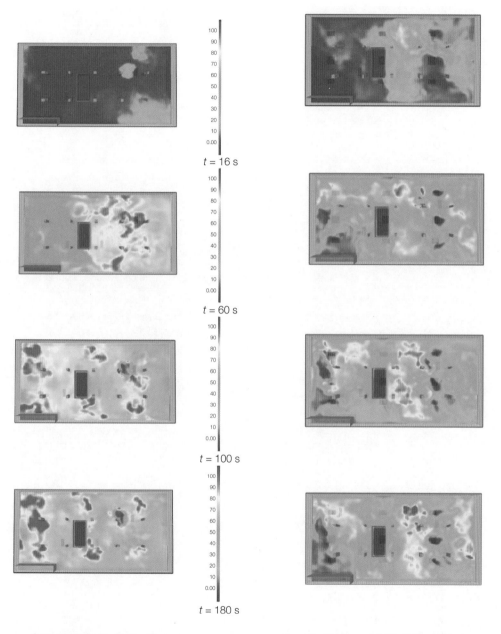

$t = 16$ s

$t = 60$ s

$t = 100$ s

$t = 180$ s

Case A: Mixing Ventilation System

Case B : Displacement Ventilation System

Table 5.3 List of building characteristics to be considered in PBD

Characteristics	List of examples
1. Occupancy	• Building classification; usage
2. Location	• Fire services access • Proximity to other hazard
3. Size and Shape	• Number of floors • Area of each floor • General layout
4. Structure	• Construction materials • Openings of external walls • Hidden Voids • Ventilation and air movement
5. Hazards	• General Layout : Location of hazardous materials, egress provision • Activities: Repair and maintenance, process and construction • Ignition source: smoking material, electrical equipment • Fuel sources: Amount of combustible materials, Fire behaviour properties
6. Fire Preventive and protective measures control	• Fire Initiation and development and control: limitation of ignition source • Smoke Development and Spread and control: provision of smoke barriers, natural smoke venting, mechanical smoke management. • Fire spread and impact and control: Fire resistive dampers provision, exposure protection • Fire Detections, warming and suppression: Provision of automatic and manual fire alarm and detection system • Occupant evacuation and control: Evacuation plan, egress route including fire isolation elements • Fire Services Intervention: Type of fire service available, characteristics of fire services capability and resources.
7. Management and use	• Regular inspections of preventive and protective measure • Training of occupancies
8. Maintenance	• Frequency and adequacy of maintenance regimes • Availability of repair personnel
9. Environmental conditions	• Ventilation and prevailing internal air currents • Prevailing patterns of wind
10. Value	• Capital • Community • Infrastructure
11. Other	• Environmental impact of fire • Fire fighting concern

The findings of these examples suggest that the current fire safety design recommendations can be insufficient to help the practical engineers. More work to explore the potential hazards and the effective means of protecting human life should be proceeded in this area.

6 Future Trends in Fire Services Designs and Control Measures

The fire services design through the PBD approach should not only rely on fire services technology, but also be integrated with the insights of other professional disciplines such as architecture, building services engineering, structural engineering and project management, in order to come up with a comprehensive life-long solution.

The fire services design is not only to consider extinguishing, controlling or suppressing the fire, but also to take into account the building characteristics through performance based design in individual project.

The fire safety design solution should be incorporated with a thorough evaluation, and perhaps with the support of computer simulation, to demonstrate that all control measures are in compliance with the buildings ordinance, regulatory standards and codes of practice, as mentioned in Section 2 above.

The cost of insurance should not be overlooked. Some designs may be perceived as having a higher risk and requiring higher premiums, and therefore may not be insurable at a reasonable cost.

In order to evaluate or design a performance based Fire Safety system, it is important to understand the buildings characteristics and its normal mode of functioning. The list of characteristics given in Table 5.3 is for ready reference.

6.1 Fire Preventive and Protective Measures Control

To assist in the fire safety system analysis, the International Fire Engineering Guide[13] recommends to consider the following six sub-systems (1 to 6) of fire preventive and protective measures control parameters.

1. Fire Initiation and Development Control—this is used to define design fires in the enclosure of fire origin as well as enclosure to which the fire has subsequently spread and how fire initiation and development might be controlled.

2. Smoke Development and Spread Control—this is used to analyse the development of smoke, its spread within the building, the properties of the smoke at locations of interest and how the development and spread might be controlled. This process enables estimates to be made of the times of critical events.

3. Fire Spread and Impact Control—this is used to analyse the spread of fire beyond an enclosure, the impact a fire might have on the structure and how the spread and impact might be controlled.

4. Fire Detection, Warming and Suppression—this is used to analyse detection, warming and suppression for fires. This process enables estimates to be made of times of critical events and the effectiveness of suppression.

5. Occupant Evacuation Control—this is used to analyse the evacuation of the occupants of a building. This process enables estimates to be made of the time required for occupants to reach a place of safety.

6. Fire Services Intervention—this is used to analyse the effects of the intervention activities of fire services in a fire including the effectiveness of suppression activities.

6.2 Approaches and Methods of Analysis

Having determined the non-compliant issues or the relevant specific objectives of the performance requirements, and taken into consideration the above six major control parameters, the following three approaches are to be used to analyse the groups of issues (or single issue) identified in the analysis strategy. The analysis may have to: (1) carry out in a comparative or absolute manner, (2) apply qualitative or quantitative methodologies, and (3) use deterministic or probabilistic tools[13].

6.2.1 Comparative or absolute approach

Comparative approach aims to determine whether the alternative solution is equivalent to or better than the deemed-to-safety or prescriptive design or benchmark design. The comparative approach is often referred to as an "equivalence" approach.

Absolute approach is the use of agreed acceptance design criteria to carry out the evaluation on an absolute basis. The results of the analysis of the trial design are matched against the objectives or performance requirements without comparing to the deemed-to-satisfy or prescriptive or "benchmark" designs.

6.2.2   Quantitative approach

The quantitative approach will reflect the decisions made with respect to the following adopted methodology:

- Formulas, equations and hand calculations
- Statistical studies
- Experiments with physical scale models
- Full-scale experimental test such as fire test or trial evacuations of real buildings
- Computer simulation of fire development and smoke spread
- Computer simulation of people movement

6.2.3   Deterministic or probabilistic approach

6.2.3.1   Deterministic Approach

This is based on physical relationships derived from scientific theories and empirical results. With a given set of initial boundary conditions, a deterministic methodology will always produce the same outcome.

An analysis using deterministic methods generally adopts a timeline approach where the time of occurrence of various events is calculated and compared. An example is given in the shaded box below (see Note 13).

ASET – RSET time-line approach

The Available Safe Evacuation Time (ASET), obtained from Sub-system "Smoke Development and Spread Control" or "Fire Spread and Impact Control" using acceptance criteria for tenability is compared with the Required Safe Evacuation Time (RSET) obtained from Sub-system "Occupant Evacuation Control".

For an absolute type evaluation, ASET should be greater than REST by a margin determined during the Performance-based design process (a safety factor), i.e.

$$ASET > RSET$$

This is shown in the figure below which depicts the time-line under consideration:

For a comparative type evaluation, this margin for the trial design should be the same or greater than that for the deemed-to-satisfy design or prescriptive design as stipulated on local regulation.

6.2.3.2   Probabilistic Approach

Probabilistic approaches use a variety of risk based methodologies. These methods generally assign reliabilities to the performance of various fire protection measures and assign frequencies of occurrence of events. They may analyse and combine several different scenarios as a part of the completed fire engineering evaluation of a building design.

Probabilistic methods generally require a lot of statistical data which are not always readily available and because of their complexity, it may involve time-consuming calculations.

An example of a methodology with "multiple design fire scenarios" using event trees is given in the following shaded box (see Note 13):

Probabilistic event-tree approach

A procedure for such an analysis may comprise the steps set out below:

1. Develop multiple design fire scenarios using event trees.

2. Quantify the design fire scenarios in terms of:
   - The times of occurrence of the events comprising each scenario (as for deterministic method) using the appropriate sub-system analysis; and
   - The probability of occurrence of the events

3. Estimate the consequences of each design fire scenario in terms of the expected number of deaths for a given population and for the entire design life of the building.

4. Estimate the Expected Risk-to-Life (ERL) which is the sum of the risks over all fire scenarios, where:

$$ERL = \frac{(\text{Expected number of deaths})}{(\text{Number of occupants} \times \text{design building life})}$$

5. Compare the ERL estimated with the acceptance criteria for the analysis. For acceptance, ERL estimated should be $\leq$ ERL acceptance. The value of ERL acceptance may be a specified number (an absolute type evaluation) or that of a design for the building that confirms to the deemed-to-satisfy or prescriptive provisions (a comparative-type evaluation).

7 Conclusion

Fire Services installations are highly important in modern buildings, especially for special and mega buildings having difficulties in conforming to current prescriptive codes, standards and legislative requirements in the Hong Kong Special Administrative Region (HKSAR). Inadequate fire prevention and protection, or misuse of general design criteria may lead to an uncontrollable situation and result in serious losses of properties and lives. Due to the advancement of technologies in tall building construction and other services areas, the design and use of special and mega buildings are subject to rapid changes. In order not to allow the fire safety standards to become out-of-date, the Authority continuously reviews and revises the fire safety requirements. It is realised that the statutory requirements are gradually shifting from the prescriptive approach to the performance based design approach.

A review on the current approval procedures, standards and legislative requirements, and the design approach of fire services installations in Hong Kong have been presented in this chapter. Also illustrated are the uses of computer simulation techniques to explore the fire risk statistically and dynamically. Examples are given for the application of computational fluid dynamics in studying flame and smoke propagation.

In future, the analysis of fire services design and control measures is expected to span from qualitative to quantitative methodologies, in both comparative or absolute manner, and probably with the aid of deterministic and/or probabilistic tools.

Notes

1. Fire Services Department. (1998). *Codes of Practice for Minimum Fire Service Installations and Equipment and Inspection, Testing and Maintenance and Equipment.* Hong Kong Special Administrative Region.

2. Buildings Department. (March, 1998). *Practice Note for Authorised Persons and Registered Structural Engineers: Guide to Fire Engineering Approach, Guide BD GP/ BREG/P/36.* Hong Kong Special Administrative Region.

3. Building Authority. (1996). *Code of Practice for the Provision of Means of Access for Firefighting and Rescue Purposes*. Hong Kong.

4. Buildings Department. (1996). *Code of Practice for Fire Resisting Construction*. Hong Kong.

5. Building Authority. (1996). *Codes of Practice for the Provision of Means of Escape in Case of Fire*. Hong Kong Special Administrative Region.

6. CIBSE. (1997). *CIBSE Guide E 1997: Fire Engineering*. The Chartered Institution of Building Services Engineers, London, UK.

7. Peacock, R. D., Jones, W. W., Reneke, P. A., & Forney, G. P. (August 2005). CFAST: Consolidated Model of Fire Growth and Smoke Transport (Version 6). User's Guide. *NIST Special Publication, 1042; NIST Special Publication, 1041, 89*.

8. McGrattan, K. B. (July 2004). *Fire Dynamics Simulator (Version 4): Technical Reference Guide, NIST SP 1018, NIST Special Publication 1018, 94*.

9. BS 7974—Application of Fire Safety Engineering Principles to the Design of Buildings—Code of Practice, (2002). British Standards Institute, UK.

10. SIMULEX User Manual, Evacuation Modelling Software, Integrated Environmental Solutions, Inc., 2001.

11. Chow, T. T., Lin, Z., & Tsang, C. F. (2008). Applying large eddy simulation in the study of fire and smoke spread at underground car park. *International Journal on Architectural Science, 7*(2), 35–46.

12. Lin, Z., Chow, T. T., Tsang, C. F., Fong, K. F., Chan, L. S., Shum, W. S. (2008). Effect of ventilation system on smoke and fire spread in a public transport interchange. *Fire Technology, 44*(4), 463–479.

13. *International Fire Engineering Guidelines*. (2005 edition). [ISBN 1741 614 562].

6

Skyscrapers—The Challenges to Building Electrical Installations

Building skyscrapers is never an easy task for architects and engineers. They all have to work together as a team to achieve a common goal: constructing the building on time within budget and with superior quality. Structural engineers deal with the structural design of the building, while the Mechanical, Electrical, and Plumbing engineers deal with their respective aspects. These tasks include elevator design and implementation, fire protection, electrical systems, water distribution, and heating, ventilation, and air conditioning system, etc.

The vertical span of skyscrapers makes significant energy loss in the power distribution. In this chapter, we will discuss in detail the power quality and energy loss problem with reference to modern electrical technology and installations.

Simon LAI and Peter WONG

Division of Building Science and Technology
College of Science and Engineering
City University of Hong Kong

1 Substation Design

A building is typically supplied with multiple sources of electricity, including feeds for normal supply from an electric utility company and an emergency or standby source of power such as diesel generator. With the load exceeds around 400 kVA, the local electric utility company provides the building with high voltage power supplies (11 kV in Hong Kong) which are reduced via step down transformers inside a substation. Ideally, if the power supplied from the utility to a substation is fed from multiple high voltage cables, the reliability of electric supply will be increased. A three-phase four-wire low voltage supply is necessary for a larger power requirement. In Hong Kong practice, 380 volts low voltage three-phase power supply (480 volts in the United States, 600 volts in Canada, and 400 volts or 380 volts in most of Europe and Asia) is distributed from a transformer having star-connected secondary winding, with the transformer's delta-connected HV winding being fed at 11 kV high voltage.

In standard low-rise buildings, the substations are located at or below ground level. For skyscrapers, the electric demand could reach well over 200 VA/m^2. A large portion of these loads are concentrated in plant rooms, such as lift motor rooms, motor control centre for central air-conditioning refrigeration plant, pump rooms, etc. They are located in basements, ground floor, intermediate floors and upper floors. The rest of the load, for example, lighting and small power, will be evenly distributed throughout the building. Substations at low level are insufficient due to a drastic boost in power consumption in terms of MVAs "load centre" on the upper floor levels. Above a certain height of the building, the voltage drop caused by the impedance of the supply conductors will become significant, and the supply voltage will fall below acceptable values. Besides, the copper losses can be kept to a minimum if large block of power can be distributed at HV to load centres at various locations of the building. Therefore, substations must be considered locating at the upper parts of the building. The considerations involved include impact on the esthetics of the façade, the space requirements and space constraints, the impact of the service space on adjacent spaces and the transportation of transformers to and from the substations.

In general, the ground level substations are required to be positioned at the periphery of the building, where high-rise substations are located on the mechanical services floors at the building periphery. These high-rise substations will be located close to the elevator shafts, so that transformers/equipment for the substations can be transported by a cargo lift in the public area inside the building through a slab opening and/or a vehicular access. To reduce size and weight of a three-phase transformer, three single phase transformers are employed for three-phase supply. Since high voltage cables must be fed to the transformers on the upper levels, the utility company will demand the HV cables to be kept away completely from any low voltage equipment and routed up the building in separate accessible cable duct. An access is necessary so that the cables can be secured and supported at regular intervals so as to relieve stress on the cables. Cables are run in a zigzag manner to overcome the gravitational force as well as to cater for the thermal movement of cables during variation of load current and ambient temperature[1].

Transformers, like other electromagnetic devices, produce a "hum" caused by alternating flux in the transformer core. This hum, known as "magnetostriction", is primarily produced at a fundamental frequency of twice the applied frequency. Planning of sound control becomes more important as power demands increase, and these transformers are placed closer to the load centre. Transformers should be installed as far as possible from areas where the sound could be objectionable, and should not be placed near multiple reflective surfaces or have sound-dampening pads placing between the transformer and the mounting surface.

1.1 Power Supply Network and Reliability

In practical terms, distributing power supply to skyscraper is resolved into two methods—radial and ring circuits. The use of parallel feeders between each substation will provide a more secure but expensive power supply system. Each feeder cable is capable of carrying the total system load and so by design there is a 50% redundancy in utilisation during normal operation. The system can be operated so that each feeder is carrying half-load and the bus-bar linking switches are normally open. In general, the usual practice is to operate one feeder live and keep the second feeder as a standby provision.

Another cost-effective alternative to improve supply security is to install a ring-circuit. This effectively ensures that electrical distribution is provided from two alternative directions to each part of the installation. The ring-circuit may be operated as a fully closed ring or, with a normally open point somewhere along its length, usually about half way round so as to balance the normal load in each leg of the ring.

2 Electrical Service Rooms and Cable Risers

Adequate number of electrical service rooms and spaces should be provided to support the skyscraper requirements. These spaces should also be large enough and practically located close to the point of utilisation. The location must allow for easy movement of equipment in and out of the service room and to the outside. Further, the spaces must be configured to accept the equipment they are to house and provide sufficient space for equipment maintenance. For example, the main switch room should be wide enough to provide an unimpeded access of minimum 650mm on one side of the switchboard for gaining access to the rear of the switchboard, and the minimum internal headroom should not be less than 2.1m as stated in the Code of Practice for Distribution Substation Design from CLP Power Hong Kong Ltd[2].

Service cores such as elevator shafts, electrical and telecommunications rooms, mechanical rooms and risers, garbage and linen chutes and other such utility spaces should be provided. The cores will extend the complete height of the skyscraper or they may rise to a specific level and then transfer and continue in a different location. Where such an offset occurs, a horizontal space should be adequately designed, to which the services can be transferred. The space requirements for electrical and telecommunication riser rooms and spaces are significant. These rooms will house different electrical equipment such as small power distribution boards, cable feeders and busducts, normal/ emergency lighting panels, security system equipment, voice and data distribution tracks and cabinets, building management system (BMS) panels, etc. To minimise the space demands, it is better to spread the equipment among several floors and to serve multiple floors with one piece of equipment, but this solution is not always possible for all types of equipment.

Cabling will be supplied radically from each electrical service room to the point of utilisation. The service raceway of cabling (such as cable tray, trunkings and conduits) exit the service rooms present another challenge. These rooms in the service cores may be located adjacent to an elevator shaft and riser shafts for air distribution; the service raceways of the cabling, facing a limited exit window, cannot penetrate such risers. In general, these cabling will exit into a public area and must be routed above a false ceiling. These constraints may limit the size of the raceway window to the extent that it cannot accommodate all required raceways, resulting in the need for additional services risers on each floor. Cabling limitation also results in additional service risers. The length of power distribution conductors will be limited by overall voltage drop. Horizontal telecommunication cables (category 5 and category 6) are limited to a maximum length of 90 metres to comply with acceptable standards.

The structure of skyscrapers usually includes levels with deep transfer beams which will cause interference in routing of services. The structural engineer will usually accommodate limited and minor penetrations through these structural beams. When these become numerous and large, careful coordination and planning will be required.

3 Harmonic and Electromagnetic Interference

Modern skyscrapers are filled with widely distributed single-phase power electronic loads. The loads are mainly switched mode single-phase power supply. The proliferation of these non-linear loads (i.e. computers, variable speed drives, discharge lighting, UPS, etc.) in recent years can produce significant harmonic distortion in building wiring systems.

Harmonics are currents or voltages at higher frequency than the fundamental power frequency. Harmonics are classified by number, or level, and by their "sequence". The general expression for harmonic currents in a balanced three-phase system is given by:

$$I_N \quad = \quad I\sin(\omega t)n \quad + \quad I\sin(\omega t + 240)n \quad + \quad I\sin(\omega t + 120)n$$

$$\text{neutral} \quad \text{phase L1} \quad\quad\quad \text{phase L2} \quad\quad\quad \text{phase L3}$$

where n is the harmonic order

It is clear from the above why the 3rd and all triple harmonics are zero sequence in nature and must always have a neutral to flow in or a delta in which to circulate. Furthermore, the 5th harmonic is seen to be backward rotating, i.e., negative phase sequence in nature. So the harmonic sequence is as follows:

Harmonic number:	1	2	3	4	5	6	7	8	9	10	11
Harmonic sequence:	+	-	0	+	-	0	+	-	0	+	-

It is important to note that each type of harmonic has different effects on distribution system. The chart in Table 6.1 shows the major effects.

Table 6.1 Effects of harmonic sequence

Sequence	Effects on a motor	Effects on the Power Distribution System
Positive	Creates forward-rotating magnetic field	1. Heating
Negative	Creates a reverse-rotating magnetic field	1. Heating 2. Motor problems
Zero	None	1. Heating 2. Creates current in the neutral of a 3-phase, 4 wire system

Excessive levels of harmonic current can have a very serious effect on several types of building power system equipment.

(1) Heating Effects

- Harmonics currents flowing in machines cause heating effects both in the conductors and in the iron core. In particular iron losses are proportional to the square of the frequency.

- In addition, some harmonics, notably the 5th harmonic, are negative sequence (backward rotating); these can give rise to additional losses by inducing higher-frequency currents in machine rotors. There are also higher losses due to skin effects in the winding.

- Harmonics will tend to flow into the system capacitance and this can give rise to the overloading of power-factor correction equipment if allowance has not been made for this possibility in the capacitor rating.

- Load currents, which contain higher frequency harmonics, increase the eddy current and hysteresis losses associated with the transformer operation. The final result is often an overheated transformer and shortened insulation life.

(2) Interference

- Harmonics can cause interference with communication, protection and signalling circuits due to electromagnetic induction or to the flow of ground currents.

- It should be noted that triplen harmonics are usually in the zero sequence mode and look like earth-fault currents. Thus there is the possibility of protection-relay mal-operation.

(3) Resonance

- Harmonics generated in one part of a system may give rise to resonance effects in another part. However, unless these resonances give rise to excessive voltages and currents they may not be significant.

- Some resonance can however be dangerous if the magnification is large due to high circuit Q factor or low damping, and remedial measures may be necessary.

Harmonic distortion in building wiring systems can also create the problem of electromagnetic interference (EMI) and disturbance to other sensitive equipment nearby, such as communication and computer systems. Malfunctioning of the equipment because of EMI can pose a serious threat to the safety, security, and reliability of the operation and function of the entire building. Furthermore, the increasing appreciation of electromagnetic compatibility (EMC) for the electrical system adds additional dimension to the concern of harmonics in building wiring systems, particularly since mandatory performance limits are now coming into force in many countries.

3.1 Effect of Harmonics on Power Quality

In addition to the failure of electrical equipment meeting their maximum rating and overheating, the power quality problems caused by these harmonic currents will create unexpected tripping of the circuit breakers and loss of communication data. In general, the problems involve:

- Voltage unbalance problem, e.g., motor overheating

- Voltage sag problem, e.g., the lighting is extinguishing

- Transient problem, e.g., the protection devices is tripped automatically due to overvoltage or overcurrent

- Earthing problem, e.g., the communication data is lost due to the noise interference.

Generally, these harmonic producing devices often produce only odd harmonics due to symmetry of positive half cycle and negative half cycle. Even harmonics usually occur during transient conditions or equipment malfunction. The most troublesome harmonic is the third harmonic. Theoretically, on a balanced three-phase four-wire wye connected load, each phase current is separated by 120°; the resultant current is cancelled and is therefore zero in the neutral conductor.

Under non-linear single-phase loads, the third harmonic from each phase crosses the zero line at the same time or phase position. The third harmonic adds in a common neutral conductor, but it's not just the third harmonic that does this. In fact, the third and odd multiples of the third are all zero sequenced. They are known as triplens, or triplen harmonics. It is worth to note that by the time all three phases of the third harmonics are added together, the sum is three times as large. The neutral current magnitudes experienced can be quite high at about, typically, 60–80 percent of the phase current values[3]. When the fundamental and the third order harmonic current are in phase, the resultant phase current will have a reduced peak value but the effective RMS current will exceed the fundamental effective RMS. However, if the fundamental and the third order harmonic are out of phase, the resultant peak will exceed the fundamental. In a pure sine wave, the crest factor (ratio of peak current to RMS) is 1.4. On a non-linear load, the current crest factor is pulsed much higher, and values of 5 have been recorded[4]. Current with high crest factors can cause operation of breakers with low tolerance to transient currents and inadvertent operation of peak acting breakers [5].

Excessive neutral currents due to the triplen harmonics in an electric wiring system contribute to the following problems:

(1) Neutral to earth voltage (NEV) will create common-mode noise problems; furthermore, this will increase the risk for stray voltage complaints and elevation of EMF level [6]

(2) Circulating currents flowing in transformers

(3) High voltage drop at loads

(4) Failure of neutral conductor; if the neutral fails on a rising busbar system serving single phase loads, over voltage can occur on each floor, as each single phase load is connected across two phases.

A high NEV may raise the problem of protection against indirect contact under earth fault protection due to a change of phase to earth voltage. The value of NEV depends on the earth impedance, the presence of harmonic currents and type of earthing arrangement.

The injected harmonic current due to non-linear loads is propagated to all distribution circuits in a skyscraper and leads to harmonic voltage distortion on the electrical system, especially the voltage harmonic distortion levels at high floors can be very high due to cumulative harmonic voltage drop. The evaluation of some sample buildings [7] shows that harmonic limits for the 3rd harmonics are difficult to satisfy. Even if harmonic current limits are fulfilled, it is still difficult to comply with harmonic voltage limits as referred to the IEEE Standard 519-1992 [8].

3.2 Impact of Earthing on Power Quality

Earthing system has a significant effect on power quality especially with harmonic currents, e.g. voltage swell under single phase to earth fault. Connection to the general mass of earth [9] in buildings is required for: (i) low voltage system earthing, (ii) low voltage equipment earthing, (iii) earthing of lightning protection system, and (iv) functional earthing of equipment, e.g. telecommunications and computer equipment. Low voltage system earth involves low frequency currents; it is used to protect insulation under fault condition, and act as reference potential and return path for leakage current The discharge current of a lightning strike is a surge phenomenon, and the function of the earthing for the lightning protection system is to provide a low resistance path for streamer and short circuit the protected structure of the building. Functional earth involves d.c. to high frequency currents; its applications are used as:

- Communication earth
  - as signalling path
  - reduce noise and cross-talks
  - stabilise battery potential and equipment potential
  - prevent electric shock
- Computer earth
  - reduce interference
  - stabilise equipment potential w.r.t. earth

The three system earths (TN-IT-TT) and their implementation are clearly defined in installation standards (*IEC 60364* and *IEE Wiring Regulation*, 17th Edition).

Choice of earthing system[10] may be determined by normal local practice; the four elements making up the consideration for installation are: (1) safety, (2) availability, (3) reliability, and (4) maintainability. However, the major concern is searching for optimum continuity of services. In summary, the system earth will be selected as the following:

- The IT system will be chosen for the electrical installation for continuity of services and maintenance services such as power supply of control and monitoring systems and maintenance services. In term of availability, the IT is the best solution.

- The TT system will be chosen for the electrical installation to continuity of services and no maintenance service. In term of safety and maintainability, the TT is the best solution.

- The TT system whose discrimination on tripping is easier to implement is chosen for standby power sources such as standby generator.

- The installation is not essential to continuity of service and maintenance service. And also, the network is very long. The TN-S system would prefer to this installation.

- In term of maintainability, fault tracking is fast in TN but repair time is often long. Conversely, tracking of the first fault may be more difficult for IT but repairs are quicker and less costly.

Under the impact of harmonic of harmonic currents, the system earth in TN-C connection should be avoided since the 3rd order harmonics and multiples of 3 flow in the PEN and prevent from being used as a potential reference for communicating electronic systems. Moreover, if the PEN is connected to metal structures, both these and the electric cables become sources of electromagnetic disturbance. Furthermore, if the system earth TN-C employs a multi-earthed three-phase four-wire distribution network, this configuration will raise a lot of serious problems such as the increase of risk for stray voltage complaints and elevation of EMF level (see Note 6). The TNC-S should also be avoided even though risks are smaller.

It is observed that the presence of third and the ninth harmonic components of neutral current will dominate the neutral-earth voltage (NEV) (see Note 6). A high NEV may raise a lot of serious problems such as the increase of risk for stray voltage complaints and elevation of EMF level if the system is a multi-earthed four-wire distribution network. In accordance with IEC 364 on Electrical Installation of Buildings, this arrangement of earthing is a TN-C system (i.e. the transformer neutral is earthed, the frames of the electrical loads are connected to the neutral, the letter C means the neutral & protective functions combined in one conductor. Moreover, this neutral voltage should stay low, typically less than 4 volts ac rms [11].

For Hong Kong earthing practice and requirements, every electrical system should have its own earth electrode system by which the exposed conductive parts of the system are connected to earth. In other words, all exposed conductive parts of the system should be connected by protective conductors to the main terminal of the system and that should be connected to earth electrode(s) via an earthing conductor. This means that every system is connected to be part of a TT system (i.e. the transformer neutral is earthed, the frames of the electrical loads are also connected to an earth connection). If the supply is taken direct from the power company's transformer or via underground cable, a bonding conductor is connected between his main earthing terminal and the Company's transformer earth or metallic sheaths of the service cable. Therefore, the installation will be operated as part of a TN-S system (letter S means neutral & protective functions are by separate conductors). However, during the measurement of the earth fault loop impedance or testing of the operation of protective devices, the said bonding conductor must be disconnected, i.e. the design and commissioning of the installation must assume a TT system.

It is understood that the practice of the power supply companies in Hong Kong is to provide a transformer substation earth bar, to which armouring of the HV supply cables and all metalwork within the substation is solidly bonded, i.e., transformer tanks, HV switchboards, etc. Additional connections to earth by means of earth rods installed along the HV cable trenches are provided, and these are also bonded to the cable armouring and the earth bar. The transformer neutral is then connected to the earth bar to form the system earth connection. Consequently, the above problem of neutral-earth voltage (see Notes 3) is not so serious when compares to a TN-C earthing for a building wiring system.

3.3 Frequency Effects on Functional Earth

As mentioned before, functional earthing is normally required for signal reference. This earthing provision is also called clean earthing as it is supposed to be on its own and should not be contaminated by leakage currents or noise from outside electrical environment. It is long a debatable topic on whether the functional earthing should be bonded to the equipment earthing.

Bonding the equipment earth and functional earth would upset telecommunication engineer and computer engineer for fear of electrical noise and stray voltage would be introduced into the clean earth. This problem is even worse if the system has harmonic current flow.

The most significant characteristics of power systems for sensitive electronic equipment are that they must behave in an orderly manner from dc to tens of megahertz. For building wiring system (50–2500 Hz, 1st to 50th harmonics), the impedance is mostly resistance and inductance, while at performance frequency range (1 MHz), they are mostly inductance and capacitance. At high frequency, the ac wiring system serves to attenuate the magnitude of noise.

Equipment earthing system employing long ground conductors in buildings where high frequencies are present exhibit high impedances in the frequency ranges of interest. Therefore they should not be used solely to provide a high frequency reference for sensitive equipment.

Both single-point and multiple point grounding systems, which employ long ground conductors, exhibit higher impedances at higher frequencies. Therefore functional earthing requires the existence of a structure that achieves the benefits of an equipotential bonding plane throughout the frequency range of interest, as shown below.

These equipotential bonding plane structures can achieve low impedances over large frequency ranges by providing a multitude of parallel paths among the various circuits attached to them. The combination of these paths results in very low current densities in the plane. Low current densities throughout the plane imply equally low-voltage drops throughout the plane.

Therefore functional earthing with equipotential ground plane structures provide the equipotential signal-ground means of choice, when signal frequencies range from dc to tens of megahertz. They assure that minimal voltage variances exist among the connected signal circuits and interconnected equipment.

3.4 *Zero Sequence Harmonic Mitigation*

There are different passive and active mitigation methods to suppress harmonics in the electrical distribution system[12]. Some harmonic compensating filters are effective and others are not, depending on the filter type and location in the building wiring system[13]. Compensation at branch circuit near the load equipment is better because this can reduce a lot wiring loss. However, there is spacing problem to place an equipment box in the tenant office. Moreover, the exceedingly high voltage harmonics distortion appeared in the riser is due to the cumulative current harmonics injected by the tenants, it is also uneconomical to install current harmonic mitigation devices at each tenant positions. An economical and practical solution would be to install current harmonics mitigation devices at the sub-circuit panel, i.e. locating them at meter/switch room of each floor.

Neutral current blocking filter, zigzag autotransformer grounding filter or active power filter can be used to mitigate the harmonic currents at the sub-circuit panel. A neutral current blocking filter will suppress the triplen harmonic but other harmonics tend to increase (see Notes 6) and decrease the ride-through capability of electronic equipment. Since the major current harmonic in a building wiring system is the 3rd harmonics, a parallel zigzag zero sequence passive filter is cost effective when compares with the active power filter to mitigate the 3rd current harmonic.

The basic triplen harmonic cancellation principle of zigzag zero sequence passive filter can be easily understood by means of Figure 6.1 and the following calculation.

For a turn ration N_2/N_1, the current circulating for instance in the primary winding 1 equals to $N_2/N_1 (i_1 - i_3)$. Furthermore, other equations are:

$$i_1 = I_{13k} = I \sin 3k\omega t$$
$$i_3 = I_{33k} = I \sin 3k (\omega t - 4\pi/3) = I \sin (3k\omega t - 4\pi)$$

The above equations give $N_2/N_1 (i_1 - i_3) = 0$ and the zigzag connected secondary winding acts therefore as an attenuator to harmonics of order 3k.

Based on the data obtained from site measurements from one case study[14] for the harmonic profile of loads and parameters of the cable for the wiring system, a zigzag zero sequence passive filter is installed into the power wiring system to see its effect on harmonic mitigation. After the instillation of the filter, the overall power quality becomes better and results are shown in Table 6.2.

As a result, the zero-sequence passive filter is able to reduce the current harmonics and voltage harmonics to meet with international standards and the statutory requirements on energy efficiency in a cost-effective manner.

Figure 6.1 Transformer with zigzag connected secondary and attenuation of harmonics of order 3k

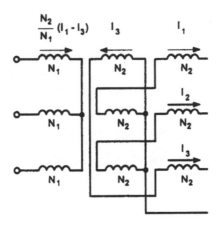

Table 6.2 Results of power quality before and after the installation of filter

	Before the installation	After the installation
ITHD	about 17–29.8%	about 15–20%
VTHD	about 5–6%	about 3.5–5%
Power Factor	about 0.77–0.85	about 0.95–0.97
I unbalance	about 15–30%	about 5–15%

3.5 Low Frequency Electromagnetic Interference

A number of ELF interference problems have been reported as the magnetic field strengths in buildings increase[15]. The most common problem reported is the computer monitor screen appears to jitter or distorted with the magnetic field exceed 1 μT (or 10 mG). In skyscrapers, substations, electrical rooms, and high-current (1000–2400 amps) risers are necessary for every 10–12 floors. Occupied areas above, below, and adjacent to the substations and busduct risers are subjected to very high (1–100 μT) and extremely high (100–10,000 μT) levels. Besides, ELF electric field will be also emanated; however, electric fields are only significantly elevated (greater than 1 kV/m) under high voltage transmission lines. Inside commercial buildings, the electric fields are normally below 50 V/m near sunshield electrical cables. Therefore, the electric field in EMF from internal 50–Hz power sources is usually not a problem inside commercial buildings.

Factors affecting magnetic filed includes: (i) load current, (ii) conductor/cable configuration/phase separation, (iii) phase sequencing of parallel cables/circuits, (iv) presence of earth ad neutral cables/conductors and the nature of neutral connections, (v) net current and (vi) the distance from the sources. These fields will create the problem of electromagnetic interference (EMI) and disturbance. The effects of EMI are:

–   malfunctioning of electronic equipment and systems

–   disturbances on power system, e.g. voltage deviation, voltage fluctuation, voltage dips

–   data security

–   mains signaling

Besides, a high level of magnetic field would also raise the concern on human safety. To prevent adverse health effects, international associations such as IEEE[16] and International Commission on Non-ionizing Protection (ICNIRP)[17] have published guidelines and standards to specify the limits of radiated electric and magnetic field. IEEE standard specifies the limit of the electromagnetic emission with frequency ranged from 3 kHz to 300 GHz. In ICNIRP guidelines, it is recommended that the limit of the electromagnetic emission with the frequency ranged from DC to 300 GHz. Therefore, ICNIRP should be referred for power frequency. The safety limit as stated in the ICNIRP is shown in Figure 6.2. The limit of magnetic field for power frequency at 50 Hz shall be below 100 μT.

If a three-phase four-wire circuit is unbalanced by more than 20%, excessive net current will flow back to the source through the neutral cable and the earth system. Closely positioned phase and neutral cables in a single phase supply system as well as phases and neutral cables in a three-phase supply system produce zero or very small magnetic fields due to the cancellation effect. However, if the net current flows through a remote route, then there is no benefit of cancellation and the resultant magnetic field is large. Any electrically conductive elements, structure or services, such as metal pipes, fences or rail track can serve as the remote route.

There are a large number of mitigation measures available[18]. The measures are basically: (i) reducing system and equipment current, (ii) increasing separation between source and affected equipment, (iii) reducing conductor spacing, (iv) re-phasing conductors, and (v) magnetic shielding.

Figure 6.2 Maximum exposure limits of magnetic field stated in ICNIRP guideline

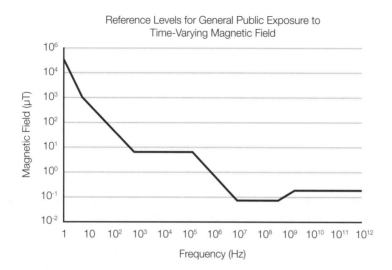

4 Lightning Protection

Most countries have a national standard or code of practice on lightning protection for buildings such as: AS1768–2007 of Australia[19], BS6651–1999 of the UK[20], the NFPA (National Fire Protection Association) code used in the USA[21], the Norm Francaise in France[22] and CP33:1996 of Singapore[23]. The International Electrotechnical Commission (IEC) has produced codes IEC 1024.1–1990 and IEC 1024.1.1–1993[24] for general use in countries that do not have their own code, or as a guide to revision for those countries that do have their code.

All national standards or codes on lightning protection have adopted the rolling sphere method of determining probable strike attachment points on the building. This method has been described by Lee[25] where a sphere radius of 45m is used, corresponding to the striking distance and collection distance for a peak lightning current of about 10kA. Based on this method, simple air terminals are required at all such possible attachment points, unless existing metallic structures will serve adequately as an air terminal. Such systems will have a large number of points on the upper parts of the building that will serve as launching points for upward streamers and connecting leaders and therefore for the attachment of the lightning channel to the building. They therefore cater for the arrival near the building of lightning leaders from all possible directions. The air terminals connect directly to down conductors that lead the lightning current to the earth electrode. This kind of protection is known as standard methods of lightning protection.

Some special purpose or proprietary interception lightning protection methods are used for some buildings. These protection systems are denoted nonstandard because they are not included in major standards for lightning protection. Some reports arguing against the effectiveness of nonstandard systems[26]. Depending on the origin of the particular device, various means for enhancing the protective property have been used including radioactive ionisation device and non-radioactive ionisation devices. All of such special purpose proprietary air terminals are claimed to provide enhanced lightning protection by causing the emission of an upward streamer will propagate towards the tip of the downward leader at an earlier stage in the attachment process than would occur for a simple air terminal of greater height than its physical height. The special purpose device is claimed to be able to attract the lightning channel to itself from a greater distance than a simple air terminal in the same position would; thus the special purpose device can compete with the simple device even though it is further from the tip of the downward leader than the simple device.

The observations of Hartono and Robiah[27], mainly based on tall buildings in Kuala Lumpur, show that neither standard nor nonstandard air terminals mounted away from the building edge are able to protect the building edge; further, the collection distances of nonstandard air terminals is not significantly different from the collection distances of simple air terminals. A conductive strip mounted along the upper outer edge of the building appears to be the essential minimum requirement for protecting skyscrapers.

5 Vertical Transportation

In order to accommodate the large population in skyscrapers, it is necessary to have a vertical transportation system that is capable of transporting all the population to their designated floor efficiently and effectively. The design would be easily done if unlimited number of lift shafts are allowed in the skyscraper. But the effective office area will then be minimised. The situation will become even worse where upper floors of a building normally deceases in floor areas. There is absolutely no room for observation deck or penthouse as all the space will be occupied by elevator machine rooms.

Sky lobby system is often used in skyscrapers to save elevator-shaft space. The building is divided vertically into sub-buildings; each with its own sky lobby floor. Large express elevators carry passengers from the ground floor to the sky lobby floors, where they can transfer to local elevator banks that take them to individual floor within the sub-buildings. This is a rather important design concept for designers of vertical transportation as skyscraper can be treated as segments of high-rise buildings[28].

The double-deck elevator system[29] was developed in 1930's to increase the transportation capacity of elevators in high-rise buildings, saving installation costs compared with adding more elevators in single-deck elevator systems. In a double-deck elevator system, two elevators cages are connected to each other; passengers at two consecutive floors can be served simultaneously, so that the handling capacity of each elevator shaft is increased. It works most efficiently in up peak traffic pattern, which stops at every other floor and serves two adjacent floors simultaneously. But it also has its drawback. Escalators are required to be installed in the entrance lift lobby to direct the passengers to the adjacent lift lobby. It is hard to control the double-deck system in other traffic pattern like regular and down-peak, due to restrictions on elevator movement caused by the fixed connection of the two decks in each elevator shaft. More efficient approaches of the double-deck elevator system are needed for its wide application since hundreds of double-deck elevators have been installed in the world during the past decades. System like destination floor guidance system can be used to optimise the problem by replacing the conventional full collective control system with up and down call buttons.

The Code of Practice for energy efficiency of lifts and escalator installations[30] set out by the Electrical and Mechanical Services Department, the Government of the Hong Kong Special Administrative Region gives some insights in designing such vertical transportation systems. The code sets out the minimum requirements for achieving energy efficient design of lift, escalator and passenger conveyor installations in buildings. The requirements can be categorised into the following areas which can help in designing vertical transportation system for skyscrapers:-

- energy efficiency;
- zoning and 5 minutes handling capacity;
- maximum allowable electrical power; and
- energy management.

Not only that each individual lift needs to be energy efficient, but that the overall lift system has to be efficient. That is, the building should have the right number of lifts serving the most efficient floor zoning arrangement (Low rise, high-rise etc.). The lifts are not being oversized and the speed relevant to the travel of the lift. With a lift system that is oversized with more lifts than are actually required, the lifts would tend to travel with a smaller load. This could mean more lift trips for the required handling capacity. It is claimed that some destination floor guidance systems are so arranged that they group passengers traveling to similar floors into the same lift car and thus reducing the number of probable stops. If it is in fact the case that probable stops are reduced, then it is possible that the number of lift movements would be reduced. This in itself reducing power consumption, but it may also lead to a reduction in the number of lifts required to satisfy the vertical transport needs of the building.

Zoning applies to skyscrapers and traction lift systems. It mainly aims at optimizing the traffic flow and reducing unnecessary start/stop cycles of the drives. One of the simple solutions is to double the capacity of the elevators, but with no increase to the size of the lift shaft. Same concept of a double decker bus, double-deck elevators are designed with two elevator cars which are attached one on top of the other. This allows passengers on two consecutive floors to be able to use the elevator simultaneously, significantly increasing the passenger capacity of an elevator shaft. Such a scheme can prove efficient in buildings where the volume of traffic would normally have a single elevator stopping at every floor. As an example, a single double-deck elevator will allow passengers to board from the ground floor and the parking floor below simultaneously. Typically, each deck will be assigned to either odd or even floors.

Architecturally, this is important as double-deck elevators occupy less building core space than traditional single-deck elevators do for the same level of traffic. In skyscrapers, this allows for much more efficient use of space, as the floor area required by elevators tends to be quite significant. In order to have double deck elevators worked efficiently, there are three important rules to be observed as listed below[31]:

- The floor heights for all floors served to be equal.
- A clear signage at the two main lift lobbies to direct passengers to the right deck.
- Suitable escalators to bring passengers to the right lobby.

For single lift installations, the cost of providing a second lift just to meet this requirement could be considered being an unnecessary cost burden, when the building operator is trying to encourage the building population to use the stairs, with the lift as an access for people with disabilities.

For lift installations where there is a bank of lifts or more than one bank of lifts, the nominated 5 minutes handling capacity of 10% of the population would be considered as minimal to maintain correct operation of the building. However, there may be projects where the developer is designing the building to suit flexi-time starting hours and may in fact be looking for handling capacities of either 10% or lower. In this case, it may well be advantageous to impose a minimum 10% handling capacity requirement. But in the lift industry in Hong Kong or similar areas, the 5 minutes handling capacity will be not less than 12% of population for high class multi-tenancy office building and not less than 11% of population for standard multi-tenancy office. A few % will have to be added if the office is a single-tenancy office building.

6 Conclusion

The design of the electrical and its associated systems for skyscrapers is very challenging compared with those for low-rise buildings. Besides the above systems, the design of the lighting treatment must work closely with a lighting design specialist when developing the power and control systems for the lighting. Aircraft warning lights are an essential requirement on any skyscraper. In the United States, these systems must comply with Federal Aviation Administration requirements. In much of the rest of the world, the International Civil Aviation Organization standards apply. The red (sometimes white) warning beacons at the top of the building and at intermediate levels should be provided as dictated by the standards. By their nature, these beacons are conspicuous, which often conflicts with the architect's vision for the building. Therefore, engineer should work together with the architect to locate these lights so that they fulfill intended purpose and limit the impact on the esthetics of the building.

Every opportunity for implementing energy saving and adopting energy efficient systems should be explored for incorporation into the design and this could facilitate improved future building operation and maintenance. Intelligent building energy management systems making use of state-of-the-art information technology and communication systems should be implemented.

Since the significant component of harmonic current for skyscraper is the third-harmonic, the most cost effective mitigation is to add a zero sequence passive filter. With the consideration of economical and practical solution, the best place is to locate the filter at the sub-circuit panel, i.e., at meter/switch room of each floor supply.

Normal local building wiring system practice has separate rising mains provided for tenant supply and landlord supply. This practice allows separate metering arrangement in electricity bill, easy in power system management for tenant and landlord supply. However, there is a problem of heavy harmonic pollution found in the rising mains for tenant supply. This harmonic pollution can be controlled in certain areas and energy loss minimisation would be effected if the rising mains for the tenant supply combined with other three phase loads (harmonic distortion tends to dilute under this kind of wiring system arrangement).

Notes

1. Yee, W., & Mak, S. (2006). High-rise substation in Skyscrapers. *The 5th Annual Power Symposium 2006*. The IET of Hong Kong, June, 2006.

2. CLP Power Hong Kong Ltd. (2008). Code of Practice No. 101 for Distribution Substation Design Version 10.0.

3. Aintablian, H. O., Hill, H. W., & Jr. (1993). Harmonic currents generated by personal computers and their effects on the distribution system neutral current. *Conf. Rec. IEEE Ind. Applicat. Society Annual Meeting*, vol. 2. (pp. 1483–1489), October 1993; Liew, A.C. (1989). Excessive neutral currents in three-phase fluorescent lighting circuits. *IEEE Trans., IA-25.4.*, pp. 776–782; IEC 555. (2001). Disturbances in supply systems caused by household and similar electrical equipment. Part 1: Definitions. Part 2 Harmonics.

4. Conroy, E. (2001). Power monitoring and harmonic problems in the modern building. *IEE Power Engineering Journal*, April, 2001.

5. Brozek, J. P. (1990). The effects of harmonics on overcurrent protection devices. *IEEE Trans., IAS.* vol.2, pp. 1965–1967.

6. Tran, T. Q., Conrad, L. E. , & Stallman, B. K. (1996). Electric shock and elevated EMF levels due to triplen harmonics. *IEEE Transactions on Power Delivery, 11*(2).

7. Du, Y., Burnett, J., Fu, Z., & Wang, L. (1997). Evaluation of harmonic limits in large office buildings. *Proceedings of IEEE Conference on APSCOM-97, Hong Kong, 1997*.

8. IEEE Standard 519-1992. IEEE recommended practices and requirements for harmonic control in electrical power system.

9. Burnett, J. (1987). Earthing and bonding in high-rise buildings. *HKIE Journal, November 1987*.

10. Lacroix, B., & Calvas, R. (1995 September). Earthing systems worldwide and evolution. *Cahier Technique*, No. 173.

11. United States Department of Agriculture. (1991 December). Effects of Electrical Voltage/Current on Farm Animals. *United States Department of Agriculture Handbook #696*. Washongton, D. C.

12. Key, T. S., & Lai, J. S. (1998). Analysis of harmonic mitigation methods for building wiring system. *IEEE Transactions on power Systems, 13*(3); Fang, Z. P. (2001). Harmonic sources and filtering approached. *IEEE Transactions on Industrial Application*, July/August. 18–25; Tain, S. L. (2000). Influence of load characteristics on the applications of passive and active harmonic filters. *Proceedings of Ninth International Conference*, Vol. 1. pp. 128–133.

13. Lai, J. S., & Key, T. S. (1997). Effectiveness of harmonic mitigation equipment for commercial office buildings. *IEEE Transactions on Industrial Application, 33*(4), 1104–1110.

14. Lai, S., Tse, N., Wong, P., & Lai, L. L. (2002). Analysis of zero sequence passive filter harmonics mitigation device for building wiring system. *Fifth International Conference on Power System Management and Control, 17–19, April 2002.* (Conference Publication No. 488).

15. Hiles, M. (1997, July–August). Magnetic field interference—Technology, economics and politics. *Power Quality Assurance*; Austin, S. (1991, September). Electromagnetic Interference in Buildings. *Building Services*; *Magnetic Field Mitigation to reduce VDU interference.* Electricity Supply Association of Australia Limited. July 1996. Australia, Melbourne.

16. IEEE C95.1–1999. (1999). Safety Levels with respect to Human Exposure to Radio Frequency Electromagnetic Fields, 3 kHz to 300 GHz. *IEEE standard*, 1999.

17. ICNIRP Safety Guideline. (1998). Guidelines for Limiting Exposure to Time-Varying Electric, Magnetic, and Electromagnetic fields (up to 300 GHz). *Health Physics, 74*(4), 494–552.

18. Du, Y., Kong, S., & Burnett, J. (2000). Management of magnetic fields in large commercial buildings. *Asia-pacific Conference on Environmental Electromagnetic CEEM 2000, May, 2000*; Vitale, L. S. (1995). *Guide to solving AC power EMF problems in commercial buildings.* Vita Tech Engineering, LLC.

19. Australian Standard, AS 1768-2007. (2007). *Lightning protection.* Standard Association of Australia.

20. British Standards Institution. (1999). *British standard code of practice for protection of structures against lightning, BS6651.*

21. National Fire Protection Association, USA. (2008). Lightning protection code. *NFPA 780.*

22. Norme Francaise (2000). *Protection contre la foudre, installations de paratonerres. NF C17-100.*

23. Singapore Standard CP33. (1996). *Code of practice for lightning protection.* Singapore Institute of Standards and Industry Research.

24. International Elctrotechnical Commission: IEC 1024.1-1990. (1990) Protection of structures against lightning, Part 1: General principles; IEC 1024.1.1-1993. (1993) Section 1: Selection of protection levels for lightning protection systems.

25. Lee, R. H. (1978). Protection zone for buildings against lightning strokes using transmission line practice. *IEEE Trans., IA–14*, 465–470.

26. Karmzyn, H., & Yeo, T. (1993). Aspects of structural lightning protection. *Lightning protection and earthing seminar, first annual technical meeting,* Centre for Management Technology, Kuala Lumpur; Mackerras, D., Darveniza, M., & Liew, A.C. (1987). Standard and non-standard lightning protection methods. *Journal of Electrical and Electronics Engineering, Australia,* 133–140; Wu, P. S., Tang H. S., Jian, X. J., Wang, S. S., & Yan, Y. J. (1989). Testing research on effectiveness of radioactive lightning conductors. *Proceedings of 6th international symposium on high voltage engineering, 1989, New Orleans, L.A.* paper 27.19.

27. Hartono, A. A., & Robiah, I. (1995). A method of identifying the lightning strike location on a structure. *Proceedings of the international conference on electromagnetic compatibility,* Kuala Lumpur. paper 4.5, pp. 112–117.

28. Siikonen, M.-L. (1998, June). Double-Deck Elevators: Savings in time and space. *Elevator World.*

29. Ishii, T. (1994, September). Elevators for skyscrapers. *IEEE Spectrum, 31(9),* 42–46; Lu. Y., Zhou, J., Mabu, S., Hirasawa, K., & Hu, J. L. (2006). A double-deck elevator group supervisory control system with destination floor guidance system using genetic network programming. *International Joint Conference, SICE-ICASE 2006, October, 2006.* pp. 5989–5994.

30. Electrical And Mechanical Services Department. (2007). *Code of Practice For Energy Efficiency Of Lifts And Escalator Installations.* The Government of the Hong Kong Special Administrative Region.

31. Sheen, P. (2006). Innovative technology to break the mold of vertical transportation system in sSkyscrapers. *The 5th Annual Power Symposium 2006.* Power & Energy Section. The IET Hong Kong, June 2006.

Reliable, Efficient and Safe Electricity Supply for Intelligent Building

Ever since the creation of the concept of Intelligent Building (IB) some 20 years ago, the advancement in technologies and business models, the need of environmental protection and sustainable development, and the appreciation of cultural, economical and geographical differences have driven the evolution of the definitions of IB across countries. In parallel with this is the development of a comprehensive assessment method to assess the "intelligence" of IB.

In this chapter, the full potential of an IB will be discussed with reference to an open building control and monitoring system approach.

Norman TSE

Division of Building Science and Technology
College of Science and Engineering
City University of Hong Kong

1 Introduction

Ever since the introduction of the concept of intelligent building (IB) 20 years ago, the definition of which has been evolving, from measuring merely the amount of intelligence the IB possessed to a more holistic approach as ". . . provides a responsive, effective and supportive intelligent environment within which the organisation can achieve its business objectives." (IBE 1991/92) [1]. New definitions of IB have since been introduced in different economical regions and countries reflecting cultural and geographical differences. In Asia, the Asian Institute of Intelligent Buildings (AIIB) established in 2000 has further defined a definition for IB [2] from the perspectives of Asian countries, described by nine "Quality Environment Modules", including:

- environmental friendliness—health and energy conservation
- space utilisation and flexibility
- human comfort
- working efficiency
- culture
- image of high technology
- safety and security measures
- construction process and structure
- cost effectiveness.

In addition to new definitions of IB, the way to realise a true intelligent building rests on the extent of integration of various building systems. Currently the level of integration would represent how intelligent a building is because that related to the exchange of information among and full integration of various building systems.

On information technology (IT) network, the current development has been on inter-connectivity of various components within a system and among systems, by the use of a wired network (copper and optical fibre) or wireless network, to achieve applications that would add values to the IB, given that IT network is everywhere and everyone affordable. The driving force behind the development of IB is therefore on the availability of a versatile and affordable IT network.

In parallel with the development of the concept of intelligent buildings is the development of methodologies for assessing the intelligence of IB. The task is not an easy one as the assessment should be based on how the "intelligence" of IB is defined, for which different cultures and countries in different stages of developments would have different interpretations. Although there has not been a universally accepted definition for IB, all proposed definitions would centre on a concept that a building must be suitable for the occupants to live in it safely, comfortably, and to work in it effectively and efficiently. As the concept of IB is originated from USA, the first definition of IB was created in USA by the Intelligent Building Institute (I.B.I.) [3]. Later in 1990s, Europe has developed its own definition of IB which is more centred on users' requirements than merely the technologies. Subsequently Asian countries have tried to develop their own definitions, such as Singapore, China and Japan. In 2001, AIIB has published a new definition of IB for Asian countries (see Note 2).

Based on the various definitions of IB reflecting varying emphasis, various IB rating methods were also derived. Up to now, there is no universally adopted single rating method. Existing rating methods are still evolving in line with the development of technologies, environmental concerns and new business models. The challenges are still in how comprehensive the rating should be without losing practicality in use.

1.1 Full Integration of IB Systems

Full integration of IB looks for the integration of electronic systems serving both the buildings operators and occupants. As electronic systems may serve purposes of various importance, such as security, safety, functionality and cost-effectiveness, integrating them into one system may raise the worry of some parties on the reliability of the integrated systems. Another hurdle of full integration is the reluctance of major market players on the adoption of an open system approach. Manufacturers prefer proprietary standards to keep customers.

As one of the key characteristics of the intelligent building is adaptability, opportunities are there to match the technologies of the building to the culture and climate of countries. This calls for a more integrated approach to the design and operation of buildings, with the appropriate use of information technology systems. Greater integration of various building and user system seems to be the way ahead.

Full integration involves inevitably a sophisticated computer and communication network, the control and monitoring of the operation of various systems would become mandatory requirements. Progress towards a fully computer-integrated building can be seen at a number of stages, as shown in Figure 7.1. (see Note 3).

Figure 7.1 Stages of integration of intelligent building

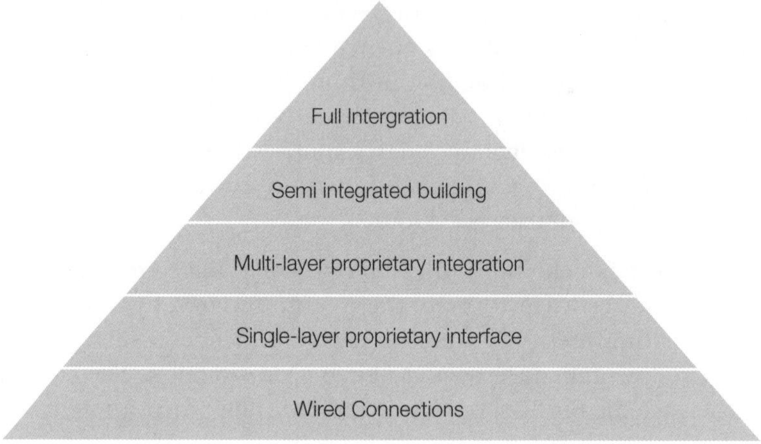

Full Intergration

Semi integrated building

Multi-layer proprietary integration

Single-layer proprietary interface

Wired Connections

1.2 Protocols for System Integration

To respond to the new market opportunities for an open system approach, companies are working on standard building control products based on LonWorks ANSI/EIA-709.1 and BACnet ISO 16484-5 protocols. Both protocols offer interoperability of products among manufacturers. At one time there were major competitions to see which protocol would win out in the industry. It is becoming clear that both protocols have been well established and manufacturers are working to offer building systems supporting both platforms.

BACnet™ is a standard for computers used in building automation and controls systems that has been developed by American Society of Heating, Refrigerating and Air-conditioning Engineers (ASHRAE). In December 1995, BACnet was also adopted by ANSI, and is now an American National Standard (ANSI/ASHRAE 135:1995)[4]. LonWorks® is a completely different solution. It is a proprietary communications technology developed by Echelon Corporation. LonWorks has been marketed for several years by the Echelon Corporation in partnership with Motorola. Various vendors have used LonWorks successfully in recent years to provide solutions for small controls systems applications, in some cases involving multiple vendors.

1.2.1   An open system based on BACnet

BACnet is specified in ANSI/ASHRAE 135:1995, "BACnet—A Data Communication Protocol for Building Automation and Control Networks". This is literally a book that describes in great detail how to create an automation and control system which can inter-operate with other BACnet systems. In BACnet terms, "inter-operate" means that two or more BACnet-speaking computer systems may share the same communications networks, and ask each other to perform various functions on a peer-to-peer basis. Although BACnet does not require every system to have equal capabilities, it is possible for designers of system components at every level of complexity to have access to functions of other automation system peers. In the BACnet world, there is no class distinction between large controllers, small controllers, sensors, actuators and operator workstations or host computers.

1.2.2   An open system based on LonWork

LonWorks is actually a family of products originally developed by the Echelon Corporation. At the core of this technology is a proprietary communications protocol called LonTalk. The term "proprietary" means that the technology was initially owned by a single proprietor—Echelon. A communications protocol is a set of rules that describe methods which can be used to manage the exchange of messages between cooperating devices that implement the protocol. The LonTalk protocol uses some advanced ideas that are unique to Echelon and their products. Because of the complexity of some of these ideas, Echelon's designers had to develop a special type of communications "chip" which was uniquely well suited to implementing LonTalk. Using this chip and the appropriate software, much of the complexity of implementing LonTalk can be absorbed completely by the communications chip, freeing the rest of the system for application tasks. This communications chip is called the Neuron®.

LonTalk looks like a very simple mailing system that provides system designers with some basic mechanisms for transporting messages between systems. LonTalk does not define what these messages contain, the sender and receiver need to agree on the content of these messages. The development of LonWorks is largely dependent on whether market players can arrive at agreement on how to define "contents" of the message.

2 Intelligent Building and Electrical Power Supply

The trend towards full integration would mean that an intelligent building must be provided with a more sophisticated, comprehensive computer and communication system. The reliable operation of the system and hence the IB depends heavily on the availability of a reliable and secure electrical power supply system. The advancement of technologies and the public awareness of environmental concerns have also posed new challenges to the design and operation of building electrical power supply systems. In general, the electrical power supply system of building must be reliable, secure, energy efficient and should provide "good quality" power. The quality of the power is "good" if it is free of power harmonics and voltage variations. The challenges faced by IB in respect to the electrical power supply are briefly discussed below.

2.1 Power Harmonics

Power harmonics are drawing attentions from researchers, engineers, and facility managers. Power harmonics are generated by nonlinear loads which include power electronics, discharging devices and highly-fluxed iron cores.[5] A load is termed as nonlinear when the current drawn by the load is not sinusoidal under a sinusoidal supply voltage. It is reported in U.S. that the percentage of nonlinear loads is around 60%, the percentage of which in Hong Kong is approaching 40%. The percentages are expected to rise due to the extensive use of digitised electrical equipment and the use of power electronics for "efficient" operation of electrical equipment. Power harmonics are causing various problems in the electrical power distribution system, among them are:

- overheating of cables and transformers;
- mechanical oscillation of rotating machines;
- capacitor bank failure;
- telecommunication interference;
- resonance at harmonic frequencies;
- inaccurate meter reading;
- false operation of protection device; and
- extra copper loss in power distribution equipment.

Harmonic voltage is a sinusoidal voltage with a frequency equal to an integer multiple of the fundamental frequency of the supply voltage.[6] Harmonic voltages can be evaluated:

1. individually by their relative amplitude (Uh) related to the fundamental voltage U1, where h is the order of the harmonic

2. globally by the total harmonic distortion factor THD, calculated using the following expression:

$$THD = \sqrt{\sum_{2}^{\infty}(U_h)^2}$$

Harmonic current is defined similarly. Harmonics of the supply voltage are caused mainly by customers' nonlinear loads connected to all voltage levels of the supply system. Harmonic currents flowing through the system impedance give rise to harmonic voltages. Although harmonic currents and system impedances and thus the harmonic voltages at the supply terminals vary in time, harmonic currents and harmonic voltages are regarded as steady state phenomenon. IEEE Standard 1159:1995 "IEEE Recommended Practice for Monitoring Electric Power Quality" sets out requirements for monitoring of electric power quality. IEEE Standard 519:1992 "IEEE Recommended Practices and Requirements for Harmonic Control in Electrical Power Systems" provides procedures for controlling harmonics on the power system along with recommended limits for customer harmonic injection and overall power system harmonic levels.

The concept of intelligent building has brought about the idea of flexible building design (refer to Note [3]). In other words, the building layout would be changed dramatically to suit new client requirement or business operation, resulting in frequent changes in load pattern and load distribution. Power quality improvement devices installed based on a particular load condition may not be suitable in other load conditions. Flexible power quality improvement schemes and harmonic mitigation approaches should be employed basing on new power conditions.

2.2 Voltage Variations

In accordance with EN50160:2000: "Voltage Characteristics of Electricity Supplied by Public Distribution Systems", voltage variations can be classified as:

1. Voltage dip—a sudden reduction of the supply voltage to a value from 90% to 1% of the declared voltage followed by a voltage recovery after a short period of time. Conventionally the duration of a voltage dip is from 10ms to 1 minute.

Voltage changes which do not reduce the supply voltage to less than 90% of the declared voltage are not considered to be dips. Voltage dip can be caused by remote power system faults.

2. Voltage swell—temporary power frequency overvoltage of relatively long duration which is usually originated from switching operations and faults such as sudden load reduction, single phase faults and non-linearities. A short duration oscillatory or non-oscillatory overvoltage is usually highly damped and with duration of a few milliseconds or less, usually caused by lightning, switching or operation of fuse.

3. Supply interruption—the voltage at the supply terminals is lower than 1% of the declared voltage. A "long interruption" is longer than 3 minutes caused by a permanent fault. A "short interruption" is up to 3 minutes caused by a transient fault. Accidental supply interruptions are unpredictable, largely random events, such as lightning strike to overhead transmission lines causing system protection operations.

As mentioned in AIIB definition (see Note 2), both electrical power quality and electromagnetic compatibility are two of the indicators specified in Index Number 1.26 and 1.66 respectively under the Green Index. The newly issued BS7671: 2008 "The 17th Edition of the IEE Wiring Regulations" which is a well known standard in electrical installations for buildings in U.K., Europe and commonwealth countries, has also put into considerations the electromagnetic influences and voltage disturbances.

2.3 Distribution Generation in Buildings

The global warming believed to be caused by the emission of carbon dioxide and the declining reserve storage of fossil fuel is drawing much attention. The seem-to-be solution to the issue is by using more renewable energy in the long run. The AIIB definition has put under Index Number 1.61, "Environmental friendliness—Substantial use of renewable energy" as one of the indicators under the Green Index. On a large scale, PV arrays for solar energy collection, wind farm, tidal wave energy, earth mass heat, etc. are being explored for as alternative energy resources. Within buildings, BIPV and micro wind turbine arrays are experimenting by researchers for powering electrical equipment in buildings. It is expected that in U.K., 10% of the total electrical power supply would be by renewables by 2010.[7] In China the plan by 2020 is to have 15% renewables not including hydro power.

The introduction of these distributed electrical power generation units into the building electrical power supply system will generally lead to an increase of the voltage magnitude experienced by the customers.[8] With highly variable sources of energy (like wind and sun) the voltage magnitude will also show a higher level of changes over a range of time scales. The use of small scale distributed generation units in building complex will change the direction of power flow and therefore the voltage drop pattern in the system. This is because the generated power is not related to the consumed power so that the total load may become negative, leading to a voltage rise in the distribution system.

Moreover the intermittent nature of the electricity supplied by wind and the sun would inevitably cause voltage variations. PV panels, some of them building integrated, requires a process of DC/AC conversion and a storage device is normally required. The process of converting DC into AC would inject current harmonics into the building electrical distribution system.

Induction generators found in many wind turbine generators has very dynamic response. The variation in the output voltage would pose problems to the stability of the output voltage. Voltage fluctuations would be an inherited property of the power generated by this energy sources.

With more and more renewable energy sources in IBs, the problems of power harmonics and voltage variations are intensified.

2.4 Energy Efficiency Requirement

Global energy reserve consumption and green house effects result in the awareness of the importance of energy conservation. As electricity is the main energy form in modern buildings, the requirements for building energy efficiency have been formulated and become mandatory in many countries, including Australia, Singapore, United Kingdom and the United States. Hong Kong is also in the process of legislating the voluntary practice in force since 1998. The voluntary Hong Kong Energy Efficiency Registration Scheme for Buildings (HKEERSB) sets minimum energy efficiency standards on four key types of fixed building services installations, namely, lighting, air-conditioning, electrical distribution and lift & escalator installations [9]. They consume up to about 80% of the total electricity consumption of a typical office building. In addition to setting limits on energy efficiency, the monitoring of energy usage is one of the key objectives of the energy efficiency codes.

Electrical energy efficiency is also one of the indicators suggested by AIIB Index Number 1.26 (see Note 2). It is expected that more stringent statutory requirements on building energy efficiency covering more building types would be in force to set limits on building energy usage, and most of them would be on the usage of electrical energy.

2.5 The Need of a Power Monitoring System

The problem of current harmonics must be monitored on a continuous basis to ensure that, over time, the incremental addition of loads does not cause excessive heating which can lead to power loss, premature failure of transformers, conductors and circuit breakers. The power harmonics mitigation measures should be adjusted in accordance with current load profiles.

As power generating facilities in future buildings would be distributed in the building electrical power distribution network, a new research interest was initiated for micro grid system for a geographic district or a building complex.[10] The need of a power monitoring system which is able to detect and record power harmonics and voltage variations and thus alerting facility management to take actions is ever more pressing.[11]

Continuous tracking of power consumption and electrical quantities provide useful data for planning future plant expansion and for ensuring that existing and future distribution feeders are adequate. It is also useful for planning the addition of loads or new alternative power supply system, such as co-generation systems and photovoltaic cells. With mandatory electrical energy efficiency requirements, the energy efficiency of an IB should be studied in both design stage and in daily operation. Evidently a power monitoring system should be installed for this purpose.

3 Reliable Electrical Power Supply Measures

Essential power supplies in the form of standby diesel generators, uninterruptible power supply unit with battery and dual power supply are commonly employed to provide a reliable electrical power supply to a building. These essential power supply systems are most suitable for IB with computer network on client server configurations. Figure 7.2 shows a combined diesel generator/UPS essential power supply system. The diesel generator is able to provide a continuous essential power supply to the loads as long as fuel refill is available. The UPS is able to provide continuous power supply to the load during diesel generator startup period which normally takes 8 seconds, and also serves as a power conditioner. The battery backup time of the UPS is normally from 5 minutes to 30 minutes.

The UPS unit may take many configurations. The one shown in Figure 7.2 is termed as a single UPS with static bypass. Other configurations such as UPS with redundancy,

Figure 7.2 Diesel generator and UPS essential power supply system

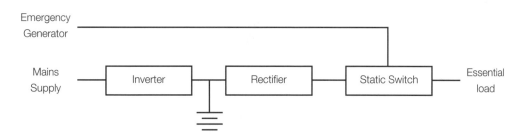

UPS with bypass, parallel UPS, etc. would be used depending on the level of power security required and the investment incurred. It is evident that the proposed essential power supply system is suitable for centralised building control systems.

With the trend of using distributed intelligence in building control systems, whereby DDC would work on standalone mode, it would be cost ineffective to provide centralised essential power supply to distributed intelligent monitoring and control units. It is therefore suggested that local battery unit would be installed to local controllers to provide short time backup power supplies.

Moreover, long period of voltage disturbances, and hence power disturbance in a well developed cities are rare. Voltage disturbance due to system faults normally would not last more than 4 cycles of waveform, in countries with 50Hz power supply, which would mean 0.08 seconds only. IT equipment would be able to ride through this short term voltage disturbance, mostly voltage dips, if they are designed to cater for it.

A standardised method of determining the robustness of IT equipment is by using the CBEMA curve, developed by the Computer Business Equipment Manufacturing Association (CBEMA) to measure the performance of all types of equipment and power systems. In 1996, a working group within CBEMA, the Information Technology Interest Group (ITIC) updated the methodology and the curve [12].

The ITIC curve plots the magnitude of the disturbance (in percentage) against the duration of the disturbance. Disturbances that fall within the envelope defined by the upper and the lower curve are typically not harmful to electrical equipment; disturbances that fall outside the envelope may disrupt or damage the equipment. Both the CBEMA and ITIC curves also plot voltage with respect to duration, but as a percentage of absolute voltage. Figure 7.3 shows the ITIC curve revised in 2000.

The ITIC curve describes an AC input voltage boundary which typically can be tolerated by most IT equipment. These curves also offer a convenient guide to define acceptable power-quality limits (see Note 12).

Figure 7.3 ITIC curve

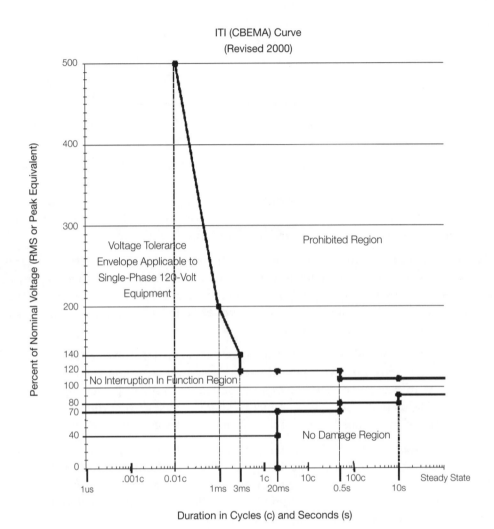

1. Steady state tolerances—an rms voltage which is either very slowly varying or is constant. The subject range is ±10% from the nominal voltage.

2. Voltage swell—having an rms amplitude of up to 120% of the rms nominal voltage, with a duration of up to 0.5 seconds.

3. Low-frequency decaying ringwave—a decaying ringwave transient with frequencies ranging from 200Hz to 5kHz. The amplitude of the transient varies from 140% for 200Hz to 200% for 5kHz.

4. High-frequency impulse and ringwave—this region deals with the amplitude and duration, rather than rms amplitude.

5. Voltage sags—sags to 80% of nominal are assumed to have a typical duration of up to 10 seconds, and sags to 70% of nominal are assumed to have a duration of up to 0.5 seconds.

6. Power interruption—may last up to 20 milliseconds.

It is important that in the course of selecting IT equipment for use as a part of the IB control and monitoring, the voltage variation tolerance of the equipment should be compliant with the ITIC curve. On the other hand, a historical record of voltage variations of an IB should be available for evaluation in respect to the ITIC curve descriptions for satisfactorily operation of IT equipment. This requires an intelligent power network monitoring system be provided in the IB, preferably as an integral part of the building control system. The intelligent power network monitoring system can be used to collect information for energy efficiency study and harmonics monitoring.

4 Intelligent Networked Power Monitoring System

In order to improve electric power quality and energy efficiency, the sources and causes of which must be known for both supply and demand sides before appropriate corrective or mitigating actions can be taken. The experience in Hong Kong is that commercially available filters and mitigation devices are theoretically useful under a prescribed power supply and distribution condition, which however may not be so in other conditions. This has resulted in significant developments in monitoring equipment that can be used to characterise disturbances and power quality variations. They are sophisticated, quite expensive and basically digital based with some memory for data storage and provisions for connection to communication network for data transfer.

For power quality issues transient in nature, they are difficult to avoid and detect unless power quality monitoring devices are installed permanently. After-fault analysis is always difficult to locate the cause of failure caused by power quality problems. Harmonic distortion and energy efficiency are classified as steady state quantities, yet they are not steady in reality and will change according to load demand pattern. The situation is even obvious on demand side such as commercial buildings where electrical loads vary in a day in relation to the business operation. With the new intelligent building design concept of flexible office layout, the office environment can be changed frequently to suit new business operation, resulting in frequent changes in load pattern.

As a result, the philosophy of power quality monitoring has evolved from a post event trouble-shooting spot check to a multitude of purposes. These purposes include: assuring compliance to statutory requirements, long term power quality monitoring for determining preventive practices, and establishing a power quality and energy efficiency profile.

The rapid changes that have taken place in the electricity industry make it of vital importance that senior facilities managers or customer services managers have easy access to processing data and information on electricity usage. Certainly power quality information is an important quantity to manage in legal terms as well as in money terms.

In accordance with the requirement of EN50160:2000 "Voltage Characteristics of Electricity Supplied by Public Distribution systems", the voltage fluctuation should be monitored for a certain period as follows:

- Power frequency   1 week
- Supple magnitude   1 week
- Flicker   1 week
- Supply voltage dips   1 year
- Short/long interruptions 1 year
- Temporary overvoltages   1 year
- Voltage unbalance   1 week
- Voltage harmonics   1 week

It is evident that spot check approach would only give some hints on the voltage quality problems; long time power monitoring is required in order to construct a comprehensive picture on voltage fluctuations.

PQ monitoring is therefore suggested to form an integral part of the intelligent control and monitoring system of the IB.

4.1 The Power Monitoring Network

The proposed system configuration[17] of the power quality monitoring network is shown in Figure 7.4. The system consists of few basic hardware components:

- PQ monitoring instrument;
- communication network; and
- central server and workstations.

Figure 7.4 Power quality system network layout

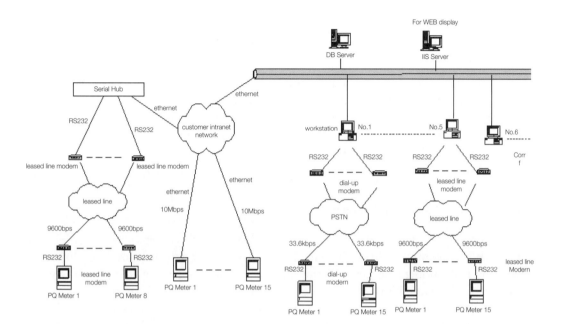

4.1.1   Power quality monitoring instrument

The power quality (PQ) monitoring instrument used is panel-mounted best provided with four voltage channels and 5 current channels which are capable of measuring 3 phase-to-neutral voltages, neutral-to-earth voltage, 3 phase, neutral and earth currents. A sampling rate of 256 samples per cycle is sufficient to catch voltage transients or current transients normally found in power distribution lines. The meter has a built-in memory sufficient to store data for 1–2 seconds temporarily. The captured data is fed to communication network through a modem. The standard outputs of the meter include:

- supply frequency;
- rms voltages and currents;
- voltage and current harmonics spectrum (up to the 64th harmonic component with a sampling rate of 256 per cycle, i.e. a sampling frequency of 12800 Hz);
- k-factor and crest factor; and

- power information such as kW, kVA, kVAr, displacement power factor and total power factor.

Phase angles at the fundamental component and at each harmonic component would also be displayed. The meter can be set to record voltage waveforms when the supply voltages exceed certain limits.

Compliance check with standards and statutory requirements can be set with the use of a standard software developed for power quality measurement purposes. The meter can be set to measure compliance with international power quality standards such as EN50160:2000 and can be configured to accommodate other industry specific standards such as IEEE519:1992 and ITIC curve.

4.1.2 Communication network

The PQ meters measure, store and transmit the data to the central system server. The data can be transmitted through leased, dial-up lines, customer intranet Ethernet network or DSL modem or through the integrated communication network of the IB, depending on the availability of communication network and the IT network of the IB. The network is specially designed to be able to use various communication means, so that different customer needs would be met. For very remote sites where communication network is not available, wireless communication using GSM format or radio can be adopted.

4.1.3 Central server and workstations

The central system consists of servers and workstations. Each workstation communicates with a group of PQ meters which are expandable.

The central system caters for the following functions:

- a server is assigned for database management where system control software and database are installed.
- a server is assigned for the internet or intranet access which is equipped with firewall.
- a workstation is assigned to store the generated reports.
- workstations are set up mainly for communicating with the PQ meters.

The configuration of the power quality system network is easily expandable by adding more PQ meters and workstations.

The system can be designed to be accessed through Web browser via internet/intranet. The advantages of using internet/intranet are well discussed by researchers for power management and monitoring[13]. For the applications in IB, the following advantages are realised:

- easy access from any locations, be it local or remote, wired or wireless;
- single point management of information;
- information sharing supported by open protocol;
- flexible solutions and scalable system configuration;
- plug and play;
- support services are available with any standard internet tools;
- user-friendly and consistent human-machine interface; and
- provide relational database access.

4.2 The Power Monitoring Central System Functions

The central system can perform many functions, such as data management, real-time fault reporting, creation of power database and routine management reporting. They are discussed below.

4.2.1 Data management facilities

Figure 7.5 is used to illustrate how data is measured and stored in the system. The system can provide data logging and storage function as well as collecting real time data. The system software will store the data into the database. The Web Java programme is used to read the required data based on the selected timestamp and then post to the Web for internet/intranet access. Also, real time data such as voltage and current are read from a VIP buffer service.

4.2.2 Real-time fault reporting function

In addition to the powerful functions of providing trending and event reports, the system should be provided with a distinct feature to be able to provide automatic real-time fault reporting functions. Depending on the type of alarms required, such as voltage sag/swell, the system would generate alarm messages such as sag/awell duration, time of the event happening, site location, to the clients/users through email and pager. It alerts them to

Figure 7.5 Data flow diagram

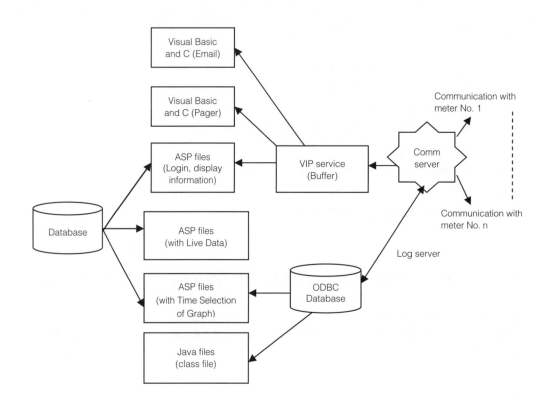

check their email or webpage for detailed analysis later. Figure 7.6 illustrates the fault alarm reporting function.

4.2.3 Power database and routine management reporting

Basically the power database for local or webpage display should be subdivided into several database sections to perform the following functions:

- trending function;
- event reporting function;
- BS EN50160 compliance reporting.

Figure 7.6 Flow chart of fault alarm reporting function

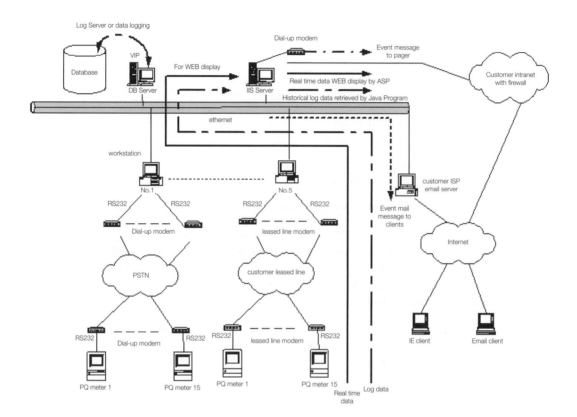

Trending function is used to store data on voltage, current, frequency and flickering, power factor, current harmonics and phase angle information. Graphical display is provided. Event function is used to store data on transient and sag/swell. Waveform, ITIC curve and 3D graph displays are available. EN50160:2000 reporting function is programmed to store data concerning voltage unbalance, voltage harmonics, interharmonics, overvoltage, voltage dips, flicker, voltage interruption, voltage magnitude and frequency. Depending on the usage, other trending can be reported as necessary.

5 Integration of Building Management System and the Power Monitoring System

For a networked power monitoring system, the requirements on processing power and data storage are very substantial such that the integration with IB system (See Note 3) is recommended to be done at high level as shown in Figure 7.7.

Typically for a networked power monitoring system for IB, the typical monitoring parameters can be grouped into four main categories:

1. Real time power information—rms values of phase & line voltages and currents, displacement and total power factors and the leakage currents.

2. Energy data—kWh, maximum KW, KVA and kVAr with time stamp.

3. Event data—minimum/maximum values with time stamp up to 1 second record for the above items (1) and (2) above, sag/swell voltage or setpoint current trigger up to half cycle calculation and transient events. In some cases, on/off/trip and busbar temperature monitoring are also included in the critical event monitor points as well.

4. Power harmonics—THD of voltage and current harmonics; individual odd current harmonic measurement may extend to the 21st harmonic order and neutral to earth voltage.

Figure 7.7 Integration of the power monitoring system

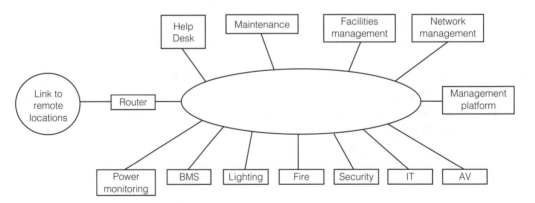

The PQ monitoring services would be accessed through Web browser. It is beneficial to link the power monitoring system to the building management system. Energy data can be linked to the building management system of the building operator such that electricity consumption can also be studied and analysed in relation to the operation of major energy using equipment, such as central air-conditioning system. The linkage can be at server level, or building management system can capture energy data directly from the PQ monitoring instrument by MODBUS or other protocols (see Note 18). MODBUS is a serial communications protocol published by Modicon in 1979 for use with its programmable logic controllers (PLCs)[14]. It has become a de facto standard communications protocol in industry, and is now the most commonly available means of connecting industrial electronic devices.

In future, event data such as transients, voltage variations and power harmonics data can also be linked to the building management system for a correlation study between power quality issues and major system operation conditions. This can be made possible by developing a new algorithm to represent event and harmonic information. Currently, Fast Forward Transform (FFT) based on Fourier Transform is commonly used to represent power harmonics. Voltage variations are normally represented by rms value variations and waveform capturing which is time-based or by statistical information. New techniques are under development that are able to represent the frequency and time information of voltage variations in a more efficient and effective means.

One of such techniques makes use of Wavelet transform [15]. The Wavelet transform is well suited for wideband signals that are not periodic and may contain both sinusoidal and impulse components as is typical of fast power system transients. In particular the Wavelet transform has the ability to focus on short-time intervals for high frequency components and long-time intervals for low-frequency components which improves the analysis of signals with localised impulses and oscillations, especially in the presence of a fundamental and low order harmonics. With the use of Wavelet transform, the sampling rate of the PQ monitoring device would be reduced, the storage capacity of the database would also be reduced, and the data transmission would be faster.

Notes

1. So, T. P., & Chan, W. K. (1999). *Intelligent building bystem.* Boston, Mass.: Kluwer Academic.

2. Wong, K. C., So, T. P., & Leung, Y. T. (2001). *The intelligent building index: IBI manual, version 2.0.* Hong Kong: Asian Institute of Intelligent Buildings.

3. Harrison, A., Joe, E., & Read, J. (1998). *Intelligent buildings in south east Asia.* London: E & FN Spon.

4. ASHRAE. (2001). BACnet—A Data Communication Protocol for Building Automation and Control Networks. *ANSI/ASHRAE 135–1995 Standard.*

5. Arrillaga, J., & Watson, N. R. (2003). *Power system harmonics.* (2nd Edition). John Wiley & Sons, Ltd.

6. British-Adopted European Standard. (2000). Voltage characteristics of electricity supplied by public distribution systems. *BSEN50160:2000.*

7. Secretary of State for Trade & Industry, United Kingdom. (2003). Energy White Paper: *Our Energy Future—Creating Low Carbon Economy.*

8. Bollen, M. H. J., & Gu, Y. H. (2006). *Signal processing of power quality disturbances.* Hoboken, NJ: Wiley-Interscience.

9. Electrical and Mechanical Services Department (EMSD) (2007). *Code of Practice for Energy Efficiency of Lighting Installations.* (2007 edition); *Code of Practice for Energy Efficiency of Electrical Installations.* (2007 edition); *Code of Practice for Energy Efficiency of Air-Conditioning Installations.* (2007 edition); Hong Kong: The Government of the Hong Kong Special Administrative Region.

10. Nigim, K. A., & Lee, W. J. (2007). Micro Grid Integration Opportunities and Challenges.*Proceedings of IEEE Power Engineering Society General Meeting 2007, Tampa, Florida, USA, Jun 2007.* CD ROM.

11. McGranaghan, M. (2001). Trends in Power Quality Monitoring. *IEEE Power Engineering Review, 21*(10), 3–9, 21; Khan, A. K. (2001). Monitoring power for the future. *Power Engineering Journal, April 2001,* IET, 81–85.

12. Technical Committee 3 (TC3) of the Information Technology Industry Council, (2000). *ITI (CBEMA) Curve Application Note.* USA: Washing DC.

13. Lee, P. K., Lai, L. L., & Tse, C. F. (2002). A web-based multi-channel power quality monitoring system for a large network. *Proceedings of the 5th International Conference on Power System Management and Control, IET UK, London, 17–19 April 2002.* pp. 112–117; Lee, P. K., Tse, C. F., & Ki, S. (2004). Application of digitized monitoring and

diagnosis system for building low-voltage power distribution system. *Proceedings of the 3rd Power Symposium 2004, IEE Hong Kong, China, June 2004*. pp. 2/1 to 2/5.

14. Modbus Organization, Inc. (2006). MODBUS Application Protocol Specification version 1.1b. USA.

15. Tse, Norman C. F. (2006). Wavelet-Based algorithm for detection of voltage fluctuation. *Proceedings of the 7th International Conference on Advances in Power System Control, Operation and Management (APSCOM) 2006, IET Hong Kong, China, 31 Oct–2 Nov 2006*, CD ROM; Tse, C. F., & Lai, L. L. (2007). Wavelet-Based algorithm for signal analysis. *EURASIP Journal on Advances in Signal Processing*, Volume 2007. Article ID 38916.

Practical Design Considerations for Grade A Office Building

The major advancement of communication, information and transportation (CIT) technologies in the past few decades poses real challenges for designers of office buildings to provide accommodation in which human beings can operate effectively.

This chapter will illustrate the design considerations of office accommodation with practical illustrations. The influences of CIT on human intelligence, work productivity and flexibility will also be discussed.

Victor LEUNG

Project Director, J. Roger Preston Ltd.

1 Adaptation to Tenancy Characteristics

Office buildings are usually designed to accommodate the maximum requirement of a range of anticipated tenancy characteristics. Naturally then, design provisions for individual tenancy areas are more often than not being over and above actual tenancy requirement. The ability to adjust or reconfigure building services systems in office buildings in accordance with actual tenancy requirement enables the buildings to serve for a variety of tenants which will increase the value of the buildings. Moreover, this will often bring forth improvement to energy-efficient building operation which is an important aspect of sustainability.

1.1 Fresh Air Requirement

The quantity of fresh air provided in office accommodation to dilute indoor air pollutants is still largely based on the research of physical characteristics of human body in the Western countries. However, Asians generally have lower metabolic rate as compared with Westerners, and accordingly should generate smaller amount of air pollutants. In addition, variations in occupant density and activities between different types of tenancy will also affect the quantity of air pollutants generated by the occupants. For instance, a corporate headquarter office may have substantially lower occupant density and intensity activity as compared with other types of office and accordingly will have less fresh air requirement. From another perspective, buildings located in an urban area with polluted atmosphere will require more fresh air provision and the level of atmospheric pollution may vary during the course of a building's life cycle.

From the above, it can be visualised that there will be ample chances in which fresh air provision for individual area does not match with actual requirement. It would be desirable to have the facility adjusting fresh air provision for different areas by means of, for instance, manual dampers or automatic constant air volume device. This would enable the facility manager to carry out periodic adjustment to optimise between energy efficiency and indoor air quality.

1.2 Cooling Requirement

The cooling capacity of an air conditioning system design required for individual area is designed based on the estimation of anticipated maximum occupant and office appliance heat generation for a range of intended types of tenant. Whilst the overall building cooling load may be in line with design capacity, individual area cooling capacity may deviate from actual requirement. In addition to possible inefficient use of energy due to unnecessary low space temperature, this over provision of cooling capacity may reduce the controllability of air conditioning system which will lead to occupant discomfort caused by excessive space temperature fluctuation or inability to maintain desirable space temperature. This is particularly the case for a variable air volume air conditioning system which actually has smaller control range than many would assume. For similar system, it would be desirable to provide manual damper in conjunction with VAV units to help increase the controllability of the units within the actual range of cooling load. Similarly, it will be desirable to provide regulating valve in conjunction with fan coil unit chilled water control valve to increase its controllability.

1.3 Power Supply Requirement

Like the above cases, power supply capacity designed for each individual tenant space is based on anticipated maximum demand and usually exceeds actual requirement. On the other hand, there may be occasions that tenant's requirements exceed power supply allowance, particularly when there is an increasing use of office equipment within a tenant's accommodation, such as the heavy use of workstations and IT equipments or the inclusion of data centre. The capability to transfer surplus power supply from an individual tenant space to those in deficit will help accommodate a variety of tenant requirements. Provisions to facilitate this include tenant power supply riser with generous capacity, spatial provision for upgrading tenant tee-off circuit breaker, adequate loading monitoring stations in the power distribution system. Spatial allowance for future installation of additional transformers and power distribution risers may be desirable in some circumstances.

2 Adaptation to Operating Conditions

The operation pattern of an office building is influenced by business activities within and around the building. It often differs noticeably from the design basis and may change from time to time during the course of the building service life cycle. Re-commissioning or modification of individual systems to bring it in line with prevalent operating condition often brings forth improved energy efficiency and operation performance.

2.1 Carpark Ventilation

The carpark ventilation air quantity required to maintain proper indoor air quality depends on the intensity of vehicle traffic inside the carpark. This in turn depends on a number of factors including occupancy rate of the building, commercial activities within and adjacency to the building, availability of nearby carpark and public transport facility; all of which affect carpark utilisation rate. The relative proportion of longstay/ shortstay tenancy and commercial activities will also affect carpark usage pattern and hence intensity of vehicle traffic inside carpark. In view of these factors, the intensity and pattern of carpark usage may vary during the course of building life cycle, and in fact vary at different times of the day and at different days of the week. Accordingly, it would be desirable for a carpark ventilation system to be equipped with air quality monitoring and modulating facility (such as variable speed fan drive) to allow the manual or automatic modulation of ventilation air quantity. There are instances which indicate that the associated improvement in energy efficiency is phenomenal.

2.2 Security System

The necessary provision of monitoring stations for security management system within a building such as closed circuit camera, patrol station, and door contact depends on the particular security management focus, which in turn will be affected by the activities both within and adjacent to the building. Security management focus becomes more apparent only upon actual operation of the building and may change in accordance with building operation pattern, short term activities or specific tenant requirement. Change of security management system focus often necessitates modification of security monitoring stations. It would be desirable if the system facilitates convenient modifications such as the use of intelligent loop for ready plug-and-play modification. In this manner, security management within a building can be efficiently brought in line with prevalent requirement and at an optimum cost.

3 Adaptation to Technological Change

The cost of various components of building operation relative to each other may change during the course of building life cycle, just as the cost for energy and human resource increase substantially relative to the cost for automation. This may be due to the effect of economic cycle, depletion of certain natural resource or evolution of technology. As a result, an optimum design solution may not remain so during the course of the life cycle of a building. Building design should consider to have appropriate provisions to facilitate implementation of alternative optimum solutions as and when available.

3.1 Indoor Air Quality

Traditionally, indoor air quality is being maintained by providing sufficient amount of outdoor air to dilute indoor generated pollutants and using appropriate filtration to remove air pollutants. However, outdoor air is becoming ineffective as a means to achieve desired indoor air quality due to increasingly polluted outdoor air. Moreover, treatment of outdoor air for use as fresh air supply in an air conditioning system accounts for a significant proportion of office building consumption which is undesirable with respect to energy cost and environmental protection. Likewise, common air conditioning system filtration also has shortcomings in achieving indoor air quality expectations as it is incapable of removing certain health affecting air pollutants such as volatile organic compound, residual suspended particles and bacteria, etc. As a result, there is now a change of air treatment concept towards eradication of pollutants such as using ultraviolet light, ionisation, photocatalytic and zeolite air purifiers in conjunction with or partially replacing traditional concepts of dilution and filtration. While these are evolving technologies with varying degrees of market acceptance, provisions (such as spatial provision in supply or return air path) to facilitate future implementation of such should be considered.

3.2 Chiller Plant

Substantially improved chiller operation efficiency in recent years has noticeably brought economic viability for chiller replacement in advance of chiller service life expiration. In addition, technologies such as electromagnetic tube cleaning device or condenser automatic ball cleaning system have become viable options both economically and technically for improving chiller operating efficiency. Chiller plant design should have foresight planning to facilitate chiller replacement or be able to implement forthcoming technologies to capture the financial opportunity when available. This includes, for example, ready delivery route and appropriate plant/piping arrangement to facilitate chiller replacement or other necessary modification work without major interruption of system or building operation. Omissions of such could be technical or financial hindrance to benefitting from financial opportunities as aforementioned.

3.3 Building Automation

In recent decades, the cost of human resource and energy has substantially increased relative to the cost of building automation system. Along with the decreasing cost of building automation system, new technologies are being launched on the market with improved reliability or effectiveness and with reduced maintenance cost. Hence, building automation solutions which have not been financially attractive is rapidly becoming economical viable, and the notable illustration is automatic modulate of pump or fan operation with variable speed drive. Other than benefiting energy efficiency, such evolving technologies also provide the potential for improvement of tenancy service and flexibility for building reconfiguration.

To capture the benefit of such evolving technologies, there is a need for building automation to be capable of accepting components from various manufacturers. For instance, systems with distributed standalone intelligence modules and open protocol system communication platform facilitate staged replacement with components from different manufacturers without disruption of system operation. Up to recent years, a number of building automation systems still cannot accommodate component of other manufacturers. However, open protocol communication platform such as BACNet has been more popular with major building automation system manufacturers. This opens up a much wider choice of components for modification or upgrading of existing systems, and introduces market competition maintenance service. Appropriate consideration of this aspect in the design will help avoid an otherwise inflexible system denying the choice of system servicing vendors during the course of building life cycle.

4 Adaptation to Business Needs

4.1 Flexible Business Practice Organisation

With the advance of information communication technology and the accelerating pace of globalisation, business practices have become much more dynamic in the past few decades in adapting to market change, with downsizing, delayering or outsourcing becoming fairly commonplace. Along with this, office core staff members are expected to be flexible in undertaking a wider range of tasks while specific business processes are being outsourced, and market focus is under change from time to time.

The increasing availability of effective communication tools such as mobile phones or PDAs enables the employee to enjoy a certain degree of freedom in determining the most effective working pattern in terms of where and when to work. This flexibility of working pattern has gained increasing acceptance by both the employees and employers; the former for quality of life and the latter for optimisation of working efficiency. This trend of employee mobility has introduced new work practices such as hot-desking/hotelling/teleworking, under which employees are given a greater freedom to choose work location within an office or even move between offices of an organisation. The possibility of adopting these new office work practices depends on the provision of intensive information communication facilities so that office layouts can be conveniently re-arranged when needed.

Lastly, office practices have been moving away from previously hierarchical differentiation between senior and lower ranking staff to emphasise on the importance of interpersonal communication. Office layout plan has been moving away from the traditional simplistic perimeter partitioned managerial office and open-plan office workstations of general staff to a variety of patterns, including managerial office migrating to internal office area and distributed partitioned meeting rooms to promote staff communication and teamwork. The capability to accommodate such a variety of patterns of office workplace layout is a challenge for office property.

All these are examples of changes of business practice which office property has to catch up with. It has been a much discussed topic that office accommodations need to provide flexibility for business practices in much the same way flexibility of employee is being expected. Accordingly, the following building service design should take into consideration:

- Lighting, fire service, and air conditioning design should accommodate as far as possible a variety of partition layouts. This is a challenge with respect to the optimisation of cost against extended built-in flexibility, and also compliance with statutory codes for various layout patterns.

- Workstation services including power supply or communication cables should be adequately provided for ready use and conveniently accessible (such as with the use of underfloor cabling) for necessary modifications.

- Utilities such as power supply and air conditioning should be adequately provided, taking into account the forthcoming information communication technology required. It will be desirable for power supply system to have provision for upgrading supply capacity for certain tenants.

- Plumbing and drainage provisions should be located at strategic locations for convenient implementation of pantry and similar facilities for tenants.

4.2 *Information and Communication Technology*

Not many decades before, business transactions were dependent on physical interpersonal contact. With the evolution of information & communication technology (ICT), business transactions even of astronomical value have been concluded by communication of partakers at remote locations and have facilitated business opportunities which were previously not viable. Information technology now enables managerial staff to conveniently carry out tasks which previously have to rely on clerical support. This has substantially simplified office setup in terms of spatial and staff requirement while at the same time increasing work efficiency. Information and communication technology have substantially changed the pattern of office work and is still gaining momentum in its already overwhelming importance in modern office workplace. It is necessary that office buildings should have adequate built-in ICT or provisions for future implementation of such technology. Related design considerations include:

- provision of plant room and/or external ventilation louvers for the installation of backup power supply/cooling system for ICT when necessary

- site provision for tenant satellite communication and other communication facilities

- space provision for communication backbone riser including provision for multi-floor tenant private pathway interfloor communication cabling

- spatial consideration for tenants to install harmonic filtering equipment to achieve necessary power supply quality for their ICT equipment

5 Practical Applications

This section provides practical applications in air conditioning design of achieving quality building services design with the following perspectives:

Reliability

With major plant/system component redundancy, simple/reliable/standalone equipment, distributed (instead of central) intelligence of control system, system/equipment maintainability, and early warning monitoring for potential system/equipment malfunctioning/failure, reliability for critical systems as follows may be achieved:

- air cooled chiller with generator power supply backup to provide stable and reliable essential chilled water supply for critical equipment

- adequate redundancy for major equipment such as chiller and chilled water pumps

- standby/dual route chilled water distribution mains

- dual air handling unit with common mains supply air duct

- comprehensive BMS monitoring to provide early warning of abnormal system operation

Indoor environment

Other than indoor comfort criteria such as space temperature, humidity, air velocity and noise levels, satisfactory indoor environmental hygiene indicated by levels of air suspended bacteria, RSP (respirable suspended particle), TVOC (total volatile organic compound) is also important since it will lead to a loss of productivity or personnel due to illness.

Optimisation of system/equipment operation

Effort is made to optimise system/equipment operation by means of appropriate system design, equipment selection, effective/integrated/properly setup automatic control system, and thorough system/equipment commissioning with a view to reduce energy and maintenance cost.

5.1 Air Conditioning Air Distribution

5.1.1 Room air diffusion

Effective air diffusion within occupied space is vital for achieving satisfactory indoor environment. Examples are uniform space temperature profile without hot or cold spot near the perimeter window and elimination of stagnant zones to avoid local accumulation of air pollutants. Air outlets should be properly selected and located with desired air quantity to achieve the favourable air diffusion index. Where necessary, mock-up room air diffusion performance test may be carried out to verify if the required air diffusion performance is being achieved with appropriate air movement velocity throughout the range of supply air quantity.

5.1.2 Dual air handling unit

Individual floor may be served by two air handling units interconnected by common mains supply air duct to allow operation of the air handling units in parallel. This arrangement has the following benefits:

- improves resiliency with the ability to maintain partial air conditioning when one unit fails;

- improves capability to serve partial load by operating only one of the two units;

- increases the overall cooling capacity since it capitalises the cooling load diversity at various locations of individual floor (e.g. different solar exposure orientation, varying occupant density/activities).

5.1.3 Packaged air handling unit

Packaged air handling units with integral motor control panel, automatic control components or acoustic treatment, etc. may be adopted to substantially replace field wiring and pipe connections with factory fabrication which is more reliable. In addition, the electrical/control equipment which may be otherwise located in plant room is now accommodated in AHU cabinet with proper access door. This will reduce the risk of damage during the course of construction and service maintenance.

The packaged arrangement in general is that components such as sensors, motorised valves, etc. are located at a more readily accessible location (as compared with traditional AHU plant room installation) which facilitates maintenance and hence reliability of the air conditioning system.

5.1.4 Air path cleanliness

The air distribution path should be designed to allow high level of air cleanliness to be maintained to protect occupant health by the following methods:

- Centrifugal fan without fan scroll (plug fan) may be adopted to facilitate cleaning of the impellor.

- AHU casing internal surface cladded with sheet metal and being constructed to minimise protrusion or recess on the casing internal surface.

- All components inside AHU and being exposed to air stream should be arranged to ascertain cleaning access.

- AHU cooling coil condensate should be minimised by supplying pre-cooled primary air to AHU to reduce bacteria growth within AHU.

- AHU drain pan may be constructed of stainless steel and designed to slope towards condensate drain outlet located at leeward side of air stream to minimise any accumulation of condensate on the drain pan.

5.2 Air Distribution Acoustic Treatment

The following are usually critical locations in acoustic analysis of air conditioning systems:

5.2.1   Directly outside AHU room

There are various major noise sources such as followings which make area directly outside AHU room usually the noisiest in an air conditioning system:

- AHU air borne noise emitted via return air inlets;

- AHU casing noise emission via AHU plant room wall;

- main supply air duct noise emission via ceiling panel.

5.2.2   Perimeter zone

Perimeter zone is normally intended for managerial personnel office accommodation and often has a lower design noise criterion, say NC35, making noise emitted from air distribution terminal units such as VAV units becoming apparent.

Combinations of various means such as the following for reducing noise emission in air conditioning systems may be considered to achieve the intended noise criteria:

- selection of low speed and high efficiency supply air fan to reduce noise generation;

- optimisation of VAV terminal unit selection for flow/inlet pressure range to reduce noise generated by excessive throttling of VAV unit control damper;

- strengthening of AHU casing to reduce noise emission via casing;

- acoustic lining in AHU fan chamber and discharge plenum to reduce air borne noise emission from AHU;

- Reinforcing main supply air duct to reduce air duct breakout noise.

With proper means to reduce noise generation in air conditioning systems, the necessity to install a silencer to attenuate noise in the main supply air duct may be reduced or even eliminated. This will help reduce energy consumption since silencers introduce noticeable air resistance.

5.3 Air Conditioning Control

VAV air conditioning system may be controlled as follows to optimise energy efficiency and equipment performance.

5.3.1   VAV AHU supply air temperature control

Variable air volume AHU supply air temperature setpoint may be reset by air flow quantity and space temperature feedback signal from VAV terminal units to maximise AHU supply air temperature whilst maintaining space temperature and individual VAV terminal unit air flow quantity within desired range. This arrangement will achieve the followings:

- Supply air quantity from individual VAV terminal unit is maintained within desired range to improve its performance in terms of stability of control.

- Necessity for operation of perimeter zone heating is reduced to achieve energy consumption.

- Chilled water supply temperature can accordingly be reset to a higher value which benefits chiller operation efficiency.

- Stuffy conditions due to inadequate air movement is being reduced.

5.3.2   VAV AHU Supply air pressure control

Variable air volume AHU supply air pressure setpoint may be reset by VAV terminal units control damper operating position feedback signal from VAV terminal units to minimise AHU supply air pressure whilst maintaining individual VAV terminal unit control damper operating position within desired range. This arrangement will achieve the followings :

- AHU supply air fan power consumption is being reduced due to AHU supply air pressure being minimised and fan operating condition at partial load is being brought closer to optimum efficiency

- Throttling at VAV terminal units are being reduced which will help to reduce air borne noise generation in addition to improving control stability

5.3.3   Return air stream carbon dioxide content

The preset AHU fresh air quantity as controlled by fresh air control damper may be reset by carbon dioxide concentration sensor located in the AHU return air stream to modulate fresh air supply quantity just enough to maintain carbon dioxide concentration level in occupied space just below allowable upper limit.

5.4 Chilled Water Distribution

Chilled water distribution to individual office may consist of:

- Normal chilled water supply which provides chilled water supply for base building air conditioning equipment such as PAU/FCU/AHU.

- Supplementary chilled water supply which serves non-essential cooling requirement of tenants. This is often taken from normal chilled water supply via open/close motorised valve and flow limiter to control chilled water supplied to individual tenant.

- Essential chilled water supply which provides reliable and constant supply temperature/pressure chilled water supply for essential air conditioning equipment of tenants such as for IT rooms. The level of resilience of essential chilled water supply may be enhanced by providing dedicated air cooled chillers and dual chilled water mains in addition to backup power supply for chilled water plant.

5.5 *Chilled Water System Control*

5.5.1 Chiller plant operation optimisation

Chillers may be staged to operate at optimum efficiency by allowing chillers to operate within a preset range of chilled water flow rate instead of traditional concept of maintaining constant chilled water flow rate. In doing so, chiller plant energy efficiency will also be increased by maximising utilisation of the heat transfer capability of the chiller condenser/evaporator and by minimising chilled/condensing water pump power consumption by reducing chiller chilled/condensing water resistance and chilled water bypass flow rate.

5.5.2 Chiller water supply temperature control

Chilled water supply temperature setpoint may be reset by chilled water temperature sensors at cooling coils to maximise chilled water supply temperature whilst maintaining temperature difference across cooling coil/heat exchanger primary circuit below preset level. This will achieve the following benefits :

- Chiller operating efficiency is improved by maximising chilled water supply temperature

- The control stability of cooling coil control valve is improved since there is less opportunity of excessively low chilled water flow rate condition

5.5.3 Chilled water supply pressure control

Chilled water supply pressure setpoint may be reset by chilled water differential pressure sensors at cooling coils to minimise chilled water supply pressure whilst maintaining cooling coil chilled water control valve opening position within preset range. This will achieve the following benefits:

- Chilled water pump power consumption is being reduced due to chilled water supply pressure being minimised and pump operating condition at partial load is being brought closer to optimum efficiency

- The control stability of cooling coil control valve is improved since there is less opportunity of excessive control valve throttling

5.6 Chilled Water Distribution System Balancing

Proper sizing of chilled water distribution regulating valves will reduce unnecessary chilled water distribution piping system resistance. Proper commissioning of chilled water distribution regulating valves will avoid cooling coil control valves being required to overcome imbalance of chilled water distribution system and hence improve their control stability.

5.7 Variable Flow Heat Exchanger

Heat exchanger heat transfer surface designed on the basis of turbulent heat transfer will have heat transfer coefficient at reduced flow rate substantially lower than that pro rata with the flow rate and hence will require flow rate to be maintained at design level even under partial load condition. In comparison, adopting heat exchanger with heat transfer surface designed with less reliance on turbulent heat transfer to achieve linear relationship between heat transfer coefficient and flow rate allows reduced flow rate under partial load condition. In so doing, equipment may be operated with increased flexibility. For instance, chillers may be operated at optimum efficiency instead of being dictated by the requirement of chilled water flow rate and plate type heat exchangers may remain in operation under partial load condition to fully utilise the available heat transfer surface, both of which will contribute to system energy efficiency.

References

Blaxter, K. (1989). *Energy metabolism in animals and man.* pp. 144. Cambridge: Cambridge University Press.

Creagh, M., & Brewster, C. (1998). Identifying good practice in flexible working.*Employee Relations, 20*(5), 490–503. MCB University Press.

Gibson, V. *Flexible working needs flexible space? Towards an alternative workplace strategy.* Department of Real Estate and Planning, The School of Business, The University of Reading. Retrieved from www.emeraldinsight.com/1463-578X.htm.

Kakihara, M., & Sorensen, C. (2004). Practising mobile professional work: tales of locational, operational, and interactional mobility. *Info, 6*(3), 180–187. Emerald Group Publishing Limited.

Parkinson, M. (1996). Flexible workforce and the role of personnel manager. *Facilities, 14,* Number 12/13.

Development of Building Services Installations in Public Housing of Hong Kong

This chapter introduces the development of building services installations in public housing of Hong Kong. From the very basic provision in 1960s to the current provision which is comparable to that of the private sector, the building services installations have undergone continual changes which are underpinned by the four core values, viz. Caring, Customer-focused, Creative, Committed (the 4 C's) of the Housing Department.

Embedded with the 4 C's, we have developed a comprehensive range of systems covering quality assurance, knowledge management and feedback review to ensure quality delivery of services. The delivery process as well as the interest of various stakeholders will also be covered.

Chi Shing HO
Chief Building Services Engineer
Housing Department

1 The Dawning of Public Housing

In 1953, the serious fire in Shek Kip Mei squatter area marked the beginning of the government's involvement in the construction of multi-storey resettlement buildings. To cope with the increasing needs of housing in the territory at that time, the then Public Works Department built Government Low-Cost Housing in the 1960s for the low income families.

With the sharp growth of population and to further cope with the demand of public housing, the Government announced the Ten-year Housing Programme in 1972. The Hong Kong Housing Authority (HA) was established in 1973 to oversee public housing issues. The Housing Department (HD) was set up in 1974 as an executive arm of the HA to discharge her statutory responsibilities which includes, inter alia, the provisions of building services installations for public housing developments.

Up to 2007, the HA has provided some 680,000 rental flats and 207,000 flats for sales accommodating more than 2.5 million people or more than one third of the total population of Hong Kong. Facilitating one third of the population to live and socialise in our housing productions has also contributed to the stability of the territory.

2 Building Services Installations in Public Housing

One of the missions of the HA is to provide affordable quality housing to low-income families who cannot afford private rental accommodations. We have two targets. First is the cost effective use of public coffer to build and second is that the housings we build must be quality housing.

In addition to providing the building fabrics that provide the shelter and environment for our tenants, building services installations are indispensable for quality living in public housing. In the early days of the 1960s, a safe and reliable lighting point and a socket that could afford tenants to light up and listen to radio already sufficed. Later on in the 1970s, socket outlets and a TV/FM outlet were provided in designated spaces in the flat for tenants to enjoy the TV and radio. Lifts were provided as the buildings increased in height. In the 1980s, adequate electricity reserve and spare MCB were reserved for subsequent installation of air-conditioners by tenants. Door phone and CCTV surveillance system was provided to the flats for sale. Emergency generator set was provided in order to comply with Fire Services Department's requirements.

In the past two decades, there were also continual improvements. To accommodate the rising use in electricity due to the growing prosperity of the territory, the electricity supply capacity for each flat had increased up to a maximum of 60A single phase. More socket outlets are located in each designated space in close collaboration with proposed furniture layout and designed usage of the flat. Our domestic buildings are designed in compliance with all relevant statutory codes and Building Energy Codes. ACVVVF[1] and DC thyristor lift control are adopted to provide more comfortable and energy efficient ride.

Nothing can be more telling of the indispensability of building services installations than the storms of complaints received after the very rare temporary outage of electricity. Tenants are not just being accommodated; they live and socialise in our housing estates for which provision of building services installations enable them to do so.

3 The Quest for Quality

3.1 The Core Values

In the context of public housing development, the missions of the HA are:

- To provide affordable quality housing and other housing related services to meet the needs of the customers in a proactive and caring manner.

- To ensure cost-effective and rational use of public resources in service delivery and allocation of housing assistance in an open and equitable manner.

To accomplish the missions, the four core values, namely, Caring, Customer-focused, Creative, Committed (the 4C's) have been acting as the cornerstones to guide the actions and the pursuit for the continuous improvements in the development of public housing.

3.2 Continuous Quality Quest

The public housing has 50 years strong history in Hong Kong and the standard of provisions has been improving along with the growing prosperity of the general public. From the very basic provision of one lighting point plus one socket outlet per flat to the nowadays full range of building services provisions which are comparable to that of private developments, the building services installations have undergone serial changes in order to meet the rising aspirations on the quality of public housing by the general public. It also sees incessant changes in the future to address the changing aspirations of the community.

3.3 Multi-Dimensional Quality Quest

With the four core values as the cornerstones, the pursuit for quality is not limited to the provision standard. It is a multi-dimensional pursuit covering reliability, cost-effectiveness, maintainability, environmental friendliness and social responsibilities as well. The pursuit for quality is not limited to end products but also includes the delivery process which takes into account the interest of workers, a wider range of relevant stakeholders and the public at large as well. As a result, open and participative delivery process is being adopted and will be further extended. Building services installations, as part and parcel of the public housing developments, have also been undergoing these changes in the delivery.

4 Current Systems in Ensuring Quality Delivery

With more than 30 years experience in the delivery of building services installations for public housing and the large number of housing flats that are produced annually, we understand that quality housing production cannot rely solely on the diligence and conscientiousness of individual staff. We instituted systems to secure the quality delivery of services across the board. In fact the then Construction Branch of the HD was the first governmental organisation in the construction field who obtained ISO 9000 certification in 1993. Working in tandem with other construction disciplines, the BS Section has developed a comprehensive range of systems as the foundation to ensure the quality of delivery. In gist the systems cover quality assurance, knowledge management and feedback.

4.1 Quality Assurance Systems

4.1.1 Quality management system

The Development & Construction Division (DCD) of the HD provides professional services for the planning, design, project management and contract administration for the construction of public housing developments under ISO 9001 certification.

In order to ensure that all quality processes are discharged effectively, the HD has developed a full range of quality manuals/guides/instructions to assist project staff. These documents provide procedural guidelines and technical standards for project teams to follow in the entire delivery process of a construction process.

4.1.2 Quality of counterparties

The HD recognises that apart from a good in-house quality management system, it is important to secure good quality counterparties to provide services for the Department. Hence, all contractors, including the Building Services contractors, are required to be registered with ISO9000 quality system.

In addition, the HD has developed an effective tool, namely Building Services Performance Assessment Scoring System (BSPASS) to assess the performance of BS contractors such that the performances of Building Services contractors are measured in an open and objective manner. The performances of building services contractors have a direct impact on the tender opportunity as well as the competitiveness in bidding for HA's contracts.

4.2 Knowledge Management (KM) Systems

In today's world knowledge is the currency. An effective knowledge management platform is pivotal to the sustainability and effectiveness of an organisation. We fully share this philosophy and we have knowledge management systems to capture the entire delivery process, from conceptual stage at drawing boards to post completion stage obtaining feedback advice from our estate management colleagues and various stakeholders. We manage our knowledge base by different means, internal meetings, cross-divisional liaison groups, dedicated information systems and IT-based collaboration platform. The valuable information is put onto the Department's intranet for sharing among staff.

4.2.1 Design review and experience sharing

Before putting into tender or adopted for use, building services designs for projects or new installation methods/details will be reviewed by experienced staff from both new works and maintenance fields. It is an open design and experience sharing platform for all ranks of building services staff to participate.

4.2.2 Building services materials

Building services materials play an important part in the overall integrity and functioning of building services installations. Knowledge and experience gained by individual staff for building services materials is shared in internal meetings. They cover the whole delivery process which includes drafting of specifications, assessment on materials, material surveillance monitoring, experience sharing with maintenance staff etc.

4.2.3 Cross-divisional liaison group

Apart from the meetings with our building services maintenance staff, we also have a cross-divisional liaison group looking after a full spectrum of issues concerning quality of construction. Where applicable, issues for building services installations can also be discussed in that platform.

4.2.4 Dedicated information system

We have a web-based application running on IT infrastructure of DCD from which users can view, browse, search and download the full content of the latest edition of Specification Library[2] as well as its subsequent updates promulgated subsequent to the launch of Specification Library.

4.2.5 IT-based collaboration platform

We have an on-line collaboration platform named HOMES[3] for HA's construction projects. HOMES is a pioneer in creating a common information backbone for the building construction industry in Hong Kong. Applying latest information technology and user-focused workflows, it provides a fast, secure and versatile information system enabling one-stop service for HA's project teams and counterparties engaged in HA's construction projects.

4.3 Feedback Systems

4.3.1 Technical feedback

As an extended development of KM, the DCD has developed the Technical Feedback System enabling the posting of technical questions or experiences to the intranet. A subject officer would study the feedback and post the recommended actions for the technical feedback. All DCD's staff can view the problem and recommend actions for the technical feedback. In this respect, the Technical Feedback System serves as an experience sharing platform and contributes to the on-going improvement to the DCD's quality system.

4.3.2 Resident survey and post-completion review

Completed projects will undergo resident survey exercise after mass intake to capture tenants' level of satisfaction/dissatisfaction on building and building services designs and provisions. The survey results will be gathered for review on the designs and provision standards for future public housing estates.

In addition, a joint post-completion review will be carried out during the maintenance period with the participation of project team, contractors, EMD's representatives from estate management and other estate stakeholders. All feedback received on the development will be raised for discussion. The participants will analyse the findings and draw conclusions which will be posted onto the DCD's intranet.

5 Looking into the Future

We have introduced the various systems in ensuring the quality delivery of building services installations. Solid foundations though they are, they are not recipes for ensuring success in the future. There will be more unprecedented challenges in development of building services installations for the future which stems from the ever increasing expectations from the community.

The general public has a rising multi-dimensional aspirations towards the development of housing. They demand us to be more caring: for the vicinity, the territory, the people and the globe. They demand us to care for the well being of people living in our development. They demand us to ensure that the globe borrowed from our future generations is well protected and sustained. They demand us to take the lead to drive new technology use. They demand us to ensure that the benefits of people affected by the development and people working for the development is being well looked after. With the core values as our cornerstones and working in tandem with stakeholders and other building disciplines of the HD, we are optimistic that we can meet the challenges.

In developing future building services installations we see the need to address the following: improved living conditions, commitment to sustainability and environmental protection, use of new technology and, last but not least, meeting social expectations. We delineate some of our recent efforts in this regard.

5.1 Improved/Better Living Conditions

5.1.1 Building designs

The HA has been adopting a decent and inexpensive design approach while striking a good balance between meeting the tenants' aspirations and the effective use of public coffer. The provision standards are continuously reviewed with reference to the feedback, survey results and project reviews captured in the KM system.

To optimise the utilisation of land resources and to address the specific situation of individual site, the HA has widely adopted site specific designs. Environmental considerations such as wind, day-light, thermal effect, traffic noise, etc. are taken into account in great depth on the building design for each project.

5.1.2 Refuse handling systems (RHS)

In order to enhance the hygienic condition of public housing estates, the HA has initiated two types of cost effective RHS for public housing estates completed in or after October 2005. These two systems, namely Central Compactor System (CCS) and Distributed Compactor System (DCS), are designed to handle refuse collection for estates of different population. The first CCS has been put in place for use in Mei Tin Estate in 2006 with good performance.

CCS is adopted for estates with daily refuse outputs of 5 tons or more. The Food and Environmental Hygiene Department (FEHD) will arrange special truck to collect the refuse container. For those estates with less population where the daily refuse outputs are less than 5 tons, DCS is used and the FEHD will collect the refuse by conventional Refuse Collection Vehicle (RCV).

Details of the CCS and DCS are introduced in Appendix 1 at the end of this chapter.

5.2 Commitment to Sustainability and Environmental Friendliness

For preserving a better community for our future generations and in tandem with the policy of government, the HA is committed to developing sustainable and environmentally friendly designs and construction of public housing estates.

5.2.1 Energy efficient installations

The HA has been a forerunner in exploring energy efficient building services installations for its public projects. Some of our recent efforts are delineated below:

- All building services installations in new public housing developments are designed to meet the relevant Energy Codes and hence registered with the EMSD as energy efficient buildings.

- By bench marking with private residential buildings, the HA manages to lower the lighting levels at lift lobbies, corridors and staircases of housing blocks, resulting in a substantial energy saving.

- To keep pace with the use of new technologies, the HA will continue to conduct trial energy efficient building services installations with a view to applying them in new housing projects in a wider scale. The previous trials include compact fluorescent lamps, photocell sensor controls, wider use of electronic ballasts, T5 fluorescent tubes, nano reflectors, LED/self-luminous signs and solar/wind powered lightings. Most of them have been included into the lighting design provisions for general applications.

- The applications of renewable energy/new technologies are being explored including installations of grid connected photovoltaic systems, micro wind turbines, photocell controlled circuitry, and study of lighting circuits controlled by digital signals transmitted through power circuits, etc. The application of grid-connected photovoltaic systems at Redevelopment of Lam Tin Estate Phases 7 and 8 is introduced in Appendix 2.

- To reduce energy loss in the electrical supply system, the power quality is reviewed on a regular basis after the intake of tenants to ensure the stability of power supply system.

- To reduce maximum demand, soft starters are adopted whenever applicable for reducing the starting current of driving motors used in plumbing installations.

- VVVF drive and high efficient gear system are being used in lift installations to reduce energy loss in driving systems.

- The use of water-cooled air conditioning systems in new commercial premises, which would result in around 10% saving in energy consumption as compared to traditional air cooled systems, is accorded a favourable consideration against other odds. Wherever possible, hybrid ventilation systems will also be adopted in air-conditioning designs to achieve higher energy saving.

5.2.2 Twin roof water tanks—An innovative approach

Twin roof water tank design was developed by the HA as a showcase of the Caring and Customer-focused core values. These core values have prompted the HA to explore possibilities of providing the tenants with a more user-friendly living environment in public housing developments.

To maintain an uninterrupted water supply during regular water tank cleansing, the twin roof water tank design was conceptualised in March 2007. The design has used twin roof water tanks to replace the traditional single roof water tank configuration.

As the name implies, the roof water tanks are divided into two with individual inlet, outlet and level control devices. When compared to the single roof water tank configuration, the twin-tank design has provided two distinct advantages during cleansing of the water tanks:

(a) uninterrupted water supply to the tenants; and

(b) elimination of possible contamination of water supply to the tenants.

Though technically feasible, implementation of such innovative design would not be made possible without the collaborative efforts between Water Supplies Department and Housing Department. The two departments have been working closely towards the common goal of achieving quality living which has led to the issue of Water Supply Department Circular Letter No. 4/2007 giving the green light to the implementation of twin roof water tank design.

Before the issue of the Circular Letter, stop valve shall not be installed at water tank inlet pipe in the case of a pumped supply. The Circular Letter has taken into account the necessity of valve installation in the water supply line in the twin roof water tank design and Clause 4.1 of the "Hong Kong Waterworks Standard Requirements" is amended to "… In the case of a pumped supply to twin cisterns, each cistern shall be fitted with an automatic control switch and a stop valve for temporary isolation purpose…".

Eastern Harbour Crossing Site Phase 4, to be completed in 2009, will be the first project completed with the twin roof tank design.

Figure 9.1 Twin roof tank schematic complete configuration

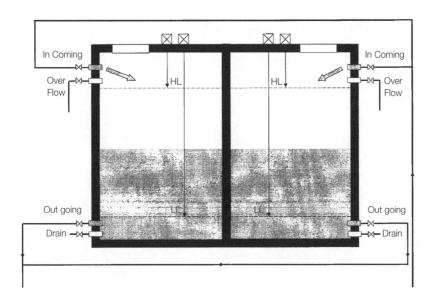

5.2.3   Hybrid ventilation system

Centralised air-conditioning system is a common installation in shopping centres in Hong Kong to provide thermal comfort to tenants. However, it consumes a large amount of energy at the same time. According to the latest energy end-use survey released by the EMSD, the energy used for space air-conditioning accounts for around 40% of the total energy consumption in retail buildings in 2004. Hence, adopting innovative method to reduce the energy consumed by air-conditioning system is essential.

In autumn and winter, the weather in Hong Kong is cool and dry and in many circumstances the demand of air-conditioning is dropped to a level that using natural ventilation is sufficient to provide pleasant thermal comfort to tenants. In view of this, hybrid ventilation system is suitable to be introduced in shopping centres. With suitable design and control algorithm and by switching the air conditioning to natural ventilation mode, it is expected that substantial energy saving can be achieved. The HA has therefore committed to install its first hybrid ventilation system in the shopping centre in Eastern Harbour Crossing Site Phase 6 which will be completed in end 2009. The design of the hybrid ventilation system is introduced in Appendix 3.

Table 9.1 EMSD latest energy end-use survey

Year	Space conditioning	Lighting	Others	Total
1994	5,253	4,784	2,784	12,821
1995	5,644	5,140	2,991	13,775
1996	5,731	5,341	3,114	14,187
1997	5,939	5,648	3,297	14,884
1998	5,855	5,661	3,307	14,824
1999	6,089	5,961	3,489	15,539
2000	6,229	6,154	3,610	15,994
2001	6,423	6,385	3,757	16,566
2002	6,292	6,255	3,680	16,227
2003	6,293	6,256	3,681	16,230
2004	6,450	6,411	3,773	16,634

Unit: Terajoule

5.3 Use of New Technologies

5.3.1   Free Wi-Fi service in public housing estates

To echo with the HKSAR Government's initiatives to provide free Wireless Fidelity (Wi-Fi) service to the public in government premises, the HA approved the provision of free Wi-Fi service to Public Rental Housing (PRH) estates in December 2007. A working team was established to liaise with the prospective information and communication technology (ICT) service providers with due consideration on the cost, method and operation of free Wi-Fi system to PRH tenants.

After negotiations and discussions with those prospective ICT service providers, it came up with a win-win arrangement that the HA would provide power supply, cable ducts and conduits, etc. to facilitate the interested ICT service providers to install and operate their Wi-Fi systems in PRH estates at their own costs. Under this arrangement, only a minimal amount of public money is invested to enable the PRH tenants to enjoy free Wi-Fi service.

The free Wi-Fi service is available to PRH tenants at the G/F entrance lobbies and designated adjoining open areas in PRH estates from 6 a.m. to 11 p.m. All the hotspots are clearly denoted by clear labels. During the operating hours, all PRH tenants could log-in the system with a unique user ID account.

In early 2008, free Wi-Fi service is provided for 120 PRH estates. The HA is keeping continuous liaison with interested ICT services providers with a view to further extending the Wi-Fi coverage and enhancing its service to PRH tenants.

5.3.2 Reception of digital terrestrial television (DTT) broadcasting

To keep in pace with the launch of DTT broadcasting, the HA had arranged the upgrading of communal aerial broadcast distribution (CABD) systems in two phases for receiving the DTT signals. The first phase of upgrading work was completed before 31 December 2007 while the second phase is scheduled to be completed by August 2008.

Benefited from the first phase upgrading work, the tenants of 68 PRH estates locating within the DTT signal receiving coverage areas of temple hill transmission station have enjoyed DTT broadcasting since 31 December 2007.

In upgrading the CABD systems, new channel amplifiers are installed for the selected DTT channels. The combined DTT and analogue TV signals are then transmitted through the existing CABD system. The DTT transmission would not affect reception of analogue television services. As the conventional terrestrial TV broadcasting will continue for at least five years after the DTT broadcasting, the PRH tenants are free to choose to enjoy their TV programmes in traditional analogue format provided by the four TV channels, or to change their TV sets to enjoy the better TV picture and sound quality from the DTT broadcast.

Different from the present analogue TV transmission, the DTT is a new method of wireless broadcasting of television programmes encoded in digitised signals. When a user receives the DTT signals from the aerial system, the digitised and compressed digital signals are decoded by a decoder (called the set-top box) so that the received DTT signals could be displayed in the tenant's TV screen. If the tenant's TV is high definition ready, the TV picture quality could be raised to as high as 1080 lines per frame. Moreover the DTT broadcasting not only improves the picture and sound quality, but also is free from problems of ghosting and signal interference.

5.4 *Meeting Social Expectation*

As a large, respectable and resourceful organisation, the general public has a high expectation on the HA. The expectation is not limited to the buildings and services it provided to tenants, but also in the delivery process of these buildings and services such that the community at large will benefit.

5.4.1 Corporate social responsibility (CSR)

CSR is the continuing commitment by business to behave ethically and contribute to economic development while improving the quality of life of the workforce and their families as well as of the local community and society at large.

CSR is closely linked with the principles of sustainable development, which advocates that enterprises should make decisions based not only on financial factors, but also on the immediate and long-term social and environmental consequences of their activities, and has increasingly been connected to corporate image, managing supply chains, partnership and stakeholder dialogue/engagement.

Since HA's construction activities may cause inconvenience and nuisance to the neighbourhood during the course of works, HA has started collaborating efforts to promote CSR during construction with the objective to create synergy in enhancing well being to the local community and promoting effective communication with the district councilors, school authorities, and the community at large.

In early 2008, the HA rolled out CSR initiatives with the contractors on a partnering spirit. The CSR initiatives include but not limit to the following:

- Greening education for adjacent schools
- Drawing display on hoarding
- Supportive services in mural wall installation
- Enhanced landscape work on site (green panels, green hoarding)
- Repair services provided for the elderly households
- Mosquito control
- Other environmental nuisance mitigation measures

5.4.2 Pay for safety and measures to enhance security of wage payment to workers

With corporate social responsibility in mind, it is the HA's conviction that a successful project can only be resulted with a good workforce. A good and trustworthy workforce must be respected and protected by the Employer.

The HA is particularly concerned with the safety and wage protection of the workers. In this end it has instituted the Pay for Site Safety, Environmental Management and Site Hygiene Scheme (PSEH) as well as the wage payment security enhancement measures.

PSEH was implemented with the warm welcome from the industry. Without lessening the obligations of the contractors, through this scheme all necessary, relevant and beneficial safety measures were treated as paid items. This provides the impetus for contractors to discharge these obligations speedily.

For workers the most important aspect for them is to get the salary they rightly deserved after their hard work. Though infrequent it was, there were cases of workers could not get the salary when their employers, various layers of contractors, went into financial difficulties. To duly protect the workers, in 2006 the HA has instituted into its contracts measures like Site Access Control system in order to clearly account for workers working in the site, wage book system to clearly log the payment records to workers employers by contractors of all tiers in project sites, and also the procurement of a labour relation officer for each project to check that all relevant wage enhancement measures were implemented. Through these measures, workers can rest assured of payment for their salary and hence can focus on delivering quality works for the HA.

Appendix 1

Refuse Handling Systems

Central Compactor System (CCS)

The system comprises the following major components:

1. A storage chamber with motorised gates connected to the bottom of refuse chute of each domestic block to correctly control the volume of refuse to be loaded to each 660-litre (660L) refuse storage bin before transporting to the refuse collection point (RCP).

2. A central refuse compactor inside RCP for receiving and compacting the refuse unloaded from the 660L storage bins to one-third of the original volume for storage in a sealed refuse storage container.

3. A sealed refuse storage container which is designed for quick loading onto the FEHD collection vehicle for transit to refuse transfer station or landfill.

Figure 9.2 Central compactor system

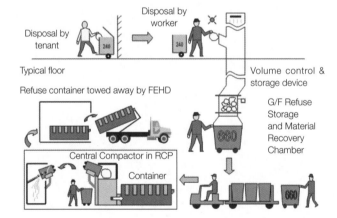

Advantages:

- Spillage during loading and transit is avoided since the bins will not be over-packed and their covers can be properly closed.

- Environmental nuisance such as noise, emission, etc. generated by the collection vehicle can be kept minimal since the time for the entire loading process is substantially reduced.

- The required RCP footprint can be reduced as compared with that required for the conventional collection process.

Distributed Compactor System

The system comprises a small-scale compactor connected to the bottom of each refuse chute to automatically compact the refuse received to half of the original volume and to partially squeeze out the foul liquid content before loading the refuse to a 660L bin for subsequent removal to the RCP for storage.

Figure 9.3 Distributed compactor system (DCS)

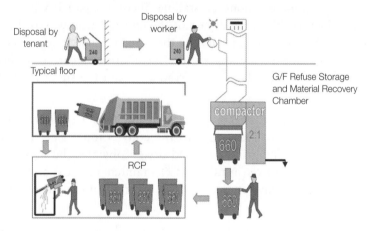

Advantages:

- This arrangement minimises odour and dirt spillage throughout the collection process.

- The compactor can also control the volume to be packed into the bin, spillage during storage and transit can also be avoided.

- The size of RCP can be reduced to create more open space due to the reduced stored volume of refuse.

- The mechanised bin cleaning device in the RCP ensures thorough cleaning of the bins before the next cycle of refuse collection.

Appendix 2

Grid-connected Photovoltaic System at Redevelopment of Lam Tin Estate Phases 7 & 8

In the interest of conservation of the environment and natural resources, the HA has decided to implement a pilot project of renewable energy technology in the form of photovoltaic (PV) system at Redevelopment of Lam Tin Estate Phases 7 & 8 (LT7 & 8). The PV system, with a capacity of 35 kW, will provide part of the electricity required for consumption by the building common facilities so as to reduce the demand of electricity. The system is scheduled to complete in early 2009.

Sunlight is directly converted to electricity by the PV system. A total of approximately 470 sq. metres mono-crystalline silicon PV panels will be installed on the roofs of three residential blocks and part of the covered walkway of LT 7 & 8. Mono-crystalline silicon PV panels are chosen for their higher energy conversion efficiency. To maximise the power gain of the system, the installation is carefully sited to take into account the available average daily solar irradiance and the shadowing effect of nearby building blocks.

The PV system will be connected directly to the power distribution network of the buildings as a power source secondary to the conventional electrical power supplied by the Power Company. The grid-connected design obviates the need of power storage batteries, thus resulting in lower installation & maintenance cost and, more importantly, elimination of the adverse environmental effect arising from the disposal of rechargeable batteries.

The electricity generated by the PV system will be conditioned and fed into the building power supply system for immediate consumption by the common facilities such as corridor lightings, lifts and water pumps, etc. The annual electrical power contributed by the system is estimated to be 43 MWh, which is around 5 to 6% of the annual electricity requirement of the common facilities in one residential block.

Through the use of PV system, the demand of electricity supply can be reduced and as a result carbon dioxide which causes greenhouse effect, produced in the electricity conversion process can be reduced. Upon the PV system of LT7 & 8 is put into operation, approximately 30 tonnes of carbon dioxide emission[4] can be reduced each year.

Appendix 3

Hybrid Ventilation System for the Shopping Centre at Eastern Harbour Crossing Site Phase 6

This appendix briefly introduces the design considerations of the hybrid ventilation system involving the following:

- Examining the external climatic conditions
- Setting internal design criteria
- Analysing the performance of hybrid ventilation system
- Designing the system
- Analysing the environmental and economical performance

External climatic conditions

Temperature and humidity conditions

Sourcing from the Hong Kong Observatory, the annual profile of the average temperature and relative humidity in Hong Kong is illustrated in the following graph.

Figure 9.4 Annual profile of the average temperature and relative humidity in Hong Kong

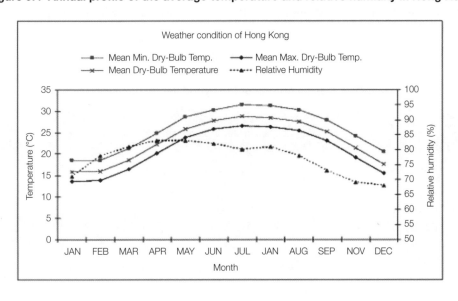

Local wind conditions

In addition to temperature and humidity, wind condition is also a critical factor that affects the effectiveness of the hybrid ventilation system. This project has made reference to the macroscopic wind condition at Kai Tak weather station (the nearest weather station) of the Hong Kong Observatory. According to the wind data, the prevailing wind direction is SE, with occasional SW occurred in July.

The microscopic wind environment around the site under different prevailing wind conditions has been analysed using CFD simulation and the results reveal that the average wind velocity at the pedestrian level and immediately around the building is in the range of 0.1 to 0.35 m/s for SE, while that of the SW wind is 0.3 to 1 m/s. In conclusion, the prevailing wind velocities in both prevailing wind directions are small to an extent that the influence to the indoor air flow rate and pattern will be very minimal.

Design Criteria

The design criteria was set by making reference to the following international standards:

- ANSI/ASHRAE Standard 55-2004, "Thermal Environmental Conditions for Human Occupancy" (pitched on 90% satisfaction level)

- ISO Standard 7730, "Moderate Thermal Environments—Determination of the PMV and PPD Indices and Specification of the Conditions for Thermal Comfort" (pitched on 10% PPD, predicted percent of dissatisfied people)

- CIBSE Application Manual AM10: 1997, "Natural ventilation in non-domestic buildings"

The design criteria of the hybrid ventilation system is summarised below:

Table 9.2 Design criteria of the hybrid ventilation system

Parameters	Summer (May to Aug)	Spring, Autumn & late summer (Mar to Apr, Sept to Oct)	Winter & late autumn (Jan to Feb, Nov to Dec)	Reference standard
Average Indoor temperature (upper limit)	27.4°C	26.6°C	25.7°C	ISO 7730
Average Indoor temperature (lower limit)	21°C	18.5°C	16°C	ANSI/ASHARE Standard 55-2004
Relative Humidity (lower/upper limit)		40%/70%		CIBSE AM10:1997

Hybrid Ventilation Performance Analysis

To assess the performance of the hybrid ventilation system, two computational simulation analyses were carried out, namely Dynamic Thermal (DTH) Analysis and Computational Fluid Dynamics (CFD) Analysis.

Dynamic Thermal (DTH) Analysis

By using DTH Analysis, the indoor thermal environment throughout a year can be simulated. The hourly values of indoor temperature and relative humidity in a year were simulated by the DTH model—IES Virtual Environment, based on the annual weather database of Hong Kong obtained from the Hong Kong Observatory and the assumptions of design parameters such as building envelop, internal loads, occupancy density, fresh air intake rate, etc.

The simulated indoor temperature and relative humidity were then compared against the design criteria mentioned above. If the simulated indoor conditions fell within the range of the design criteria, natural ventilation would be suitable to operate to provide adequate comfort level to tenants. Otherwise, free cooling or air-conditioning system should be used. The simulation revealed that the possible operating time of natural ventilation mode was around 520 hours in a year which represented around 10% of the total building operating hours (5,460 hours a year). An annual saving of cooling load of 66.7MWh was anticipated.

Computational Fluid Dynamics (CFD) Analysis

CFD Analysis could simulate the indoor air flow pattern and temperature distribution inside the building during the natural ventilation mode of hybrid ventilation system. It was also used to enhance the performance of the hybrid ventilation system through optimizing design, such as opening orientation, location and size. CFD simulation software—Star-CD was adopted for this analysis.

Four design scenarios were analysed using the CFD software. It was found that in the optimum scenario, the average air speed inside the building was within a range from 0.1m/s to 1m/s and indoor air temperature at the occupancy zone ranged from 21°C to 25°C, satisfying the design comfort criteria. No dead corner or hot spot was found and hence no additional mechanical ventilation was considered necessary.

System Operation and Monitoring

There are three operating modes for the hybrid ventilation system—natural ventilation, free cooling and air-conditioning modes—which are determined by the climate and the outdoor air conditions. For example, when the outdoor air is cool and dry enough, natural ventilation can be adopted. However, if the outdoor air is hot and humid, air-conditioning system should be turned on to provide a comfort indoor environment. The following table summarises the possible operating mode of the hybrid ventilation system in different seasons:

Table 9.3 Possible operating mode of the hybrid ventilation system

Month	Relevant Season	Description of weather conditions	Possible operating mode of hybrid ventilation
January, February, November, December	Winter and late autumn	More cloudy, with occasional cold fronts followed by dry northerly winds. It is not uncommon for temperatures to drop below 10°C in urban areas.	Mainly natural ventilation, supplement with free cooling
March & April	Spring	Pleasant, occasional spells of high humidity, fog and drizzle.	Mainly nautral ventilation & free cooling, with air conditioning in critical period
May to August	Summer	Hot and humid with occasional showers and thunderstorms; Most likely to be affected by tropical cyclones	Mainly air conditioning, with supplementary free cooling in certain hours. Possible nautral vent in night time or early morning
September & October	Autumn and late summer	Pleasant breezes, plenty of sunshine and comfortable temperatures.	Mainly natural ventilation & free cooling, supplement with air conditioning in critical time

Hybrid ventilation operation modes will be controlled and monitored by Central Control and Monitoring System (CCMS). A set of air temperature and humidity sensors will be provided at the public circulation zone of each floor to monitor the indoor conditions. Anemometer and outdoor air temperature sensors will be installed to monitor the outdoor air conditions during natural ventilation mode. Rain sensor will also be installed on the roof floor to detect the presence of rain. If rain is detected, the openable windows on upper floors of the shopping centre which are susceptible to ingress of rain will be close automatically.

In moderate climate period, natural ventilation mode will be adopted in the shopping centre and all the openings will be open. Natural ventilation will be adopted only when the indoor conditions fall within the design criteria. The indoor conditions will be sent back to the CCMS in order to determine a suitable operation mode. When outdoor air condition cannot satisfy the pre-set threshold value, free cooling or air-conditioning mode will be operated instead of natural ventilation.

Environmental and Economic Analysis

Environmental Analysis

The reduction of cooling load simulated by the dynamic thermal model was around 64MWh/year and the corresponding electricity saving per year was estimated to be 33MWh[5]. Around 23 tonnes of carbon dioxide emission can be reduced each year.

Economic Analysis

The hybrid ventilation system involves the provision of top-hung/louvred type motorised ventilation windows, CCMS interface and the associated conduit/wiring system. The estimated installation cost was estimated to be $530,000. The payback period for the hybrid ventilation system was estimated to be around 16.5 years.

Notes

1. Alternating current variable voltage variable frequency

2. Specification Library is a collection of specifications for all building construction installations in a systematic manner which facilitates users to customise the specifications for individual project.

3. Housing Construction Management Enterprise System

4. Based on an emission factor of 0.7 tonne of carbon dioxide per MWh power generation being adopted by the Environment Bureau

5. In general, an air-cooled air conditioning system consumes 1.83 kWh of electricity for 1 ton of cooling capacity.

Advanced Building Services Systems for AsiaWorld-Expo in Hong Kong

AsiaWorld-Expo, an international exhibition and events centre located at the Hong Kong International Airport, was opened in December 2005. It is one of the world's largest high-tech facilities with exhibition complex fully integrated into the airport. The total site area of AsiaWorld-Expo is around 17 hectares, gross floor area is 128,000m^2 and net column-free exhibition floor area is 66,000m^2. It consists of eight typical exhibition halls together with one large exhibition hall and one 19m high multifunction hall which can be used for concert.

We will examine the exhibition centre in detail and particularly its special features, including the utilities tunnel network, ice storage system and the textile air distribution system, etc.

Thomas Kwok Cheung CHAN

Executive Director of Parsons Brinckerhoff (Asia) Ltd.

Herbert Lung Wai LAM

Associate Director of Parsons Brinckerhoff (Asia) Ltd.

Acknowlegdements:
The authors would like to thank AsiaWorld-Expo Management Ltd.
and Parsons Brinckerhoff (Asia) Ltd. for their support to this essay.

1 Introduction

The AsiaWorld-Expo, an international exhibition and events centre located at the Hong Kong International Airport, opened on December 2005. It is one of the world's largest high-tech facilities with exhibition complex fully integrated into an international airport, making it easily accessible to the 47 million international air travellers arrived in Hong Kong each year. The AsiaWorld-Expo complex also has its own rail station, enabling visitors direct access to the Airport Express train platform which connects them to any destination served by Hong Kong MTR.

The AsiaWorld-Expo is located at the fringe of the SkyCity, northeast to the Passenger Terminal Building of the Hong Kong International Airport at Chek Lap Kok.

This chapter describes the advanced building services systems equipped in this sophisticated AsiaWorld-Expo complex to support various types of events from exhibitions to large arena-style shows such as world-class concerts.

Figure 10.1 Location plan

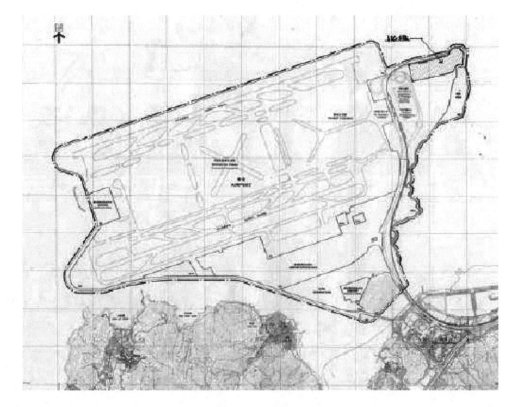

2 Building Description

The total site area of the AsiaWorld-Expo is around 17 hectares, with gross floor area 128,000m² and net column-free exhibition floor area 66,000m².

In order to cater for a wide range of events, the complex consists of several modular elements. There are altogether ten halls, including eight totally typical self-supporting, interconnecting halls with each over 5,600m² divided by full height acoustic partitions, a large (North-East) exhibition hall and a multipurpose hall with 19m clear height, providing a total of 66,000m² of column-free exhibition area. Figures 10.2 and 10.3 show the external view of the AsiaWorld-Expo complex.

Figure 10.2 External view of AsiaWorld-Expo

Figure 10.3 External view of AsiaWorld-Expo adjacent to airport

Figure 10.4 Functional zoning of the AsiaWorld-Expo complex

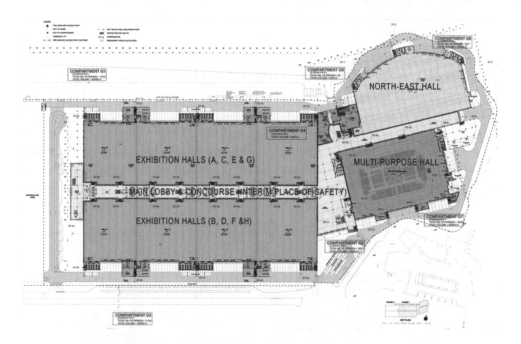

To serve as an international exhibition and events centre for various types of events, AsiaWorld-Expo has very high-ceiling exhibitions halls (10m to 19m clear headroom); some of the small exhibition halls can be inter-connected to become a large venue for international mega exhibitions. Noting that utilities to exhibition booths such as electrical power supply, telephone and telecommunication outlets, water supply and drainage are provided via floor utilities tunnel.

Hall 1 (Multi-Purpose Hall) is the biggest indoor seated entertainment arena in Hong Kong and is intended to host sophisticated world-class events ranging from superstar concerts to top sports and entertainment extravaganzas. It can be transformed into a large concert or sporting arena (Figure 10.5), using movable seating to achieve a variety of configurations, ranging up to 13,500 seats.

Figure 10.5 Exhibition hall converted to concert show

③ Main Plants and Systems Overview

Major building services systems provided in the AsiaWorld-Expo complex are briefly described in the following sections which include air-conditioning and mechanical ventilation, electrical services, plumbing and drainage, fire services, gas supply and extra low voltage systems. Locations of ACMV and Electrical plant rooms are shown in Figure 10.6 and Figures 10.11 & 10.12 respectively for reference.

3.1 Air-Conditioning and Mechanical Ventilation (ACMV) System

The AsiaWorld-Expo complex is served by a centralised chiller plant with the cooling capacity of around 5,044RT plus ice-storage by means of ice ball technology. Refrigerant R134a with zero ozone depletion potential is used as a cooling medium for environmental protection and to comply with the "Montreal Protocol" for the phasing out of the use of chlorofluorocarbons chemical products.

The halls and concourses are served by air handling units for enhancing the air distribution/movement and for minimising the future maintenance works. AHUs are located at each hall to minimise the duct run with smaller fan duty. Fan coil units with

Figure 10.6 Location of ACMV plant rooms in AsiaWorld-Expo

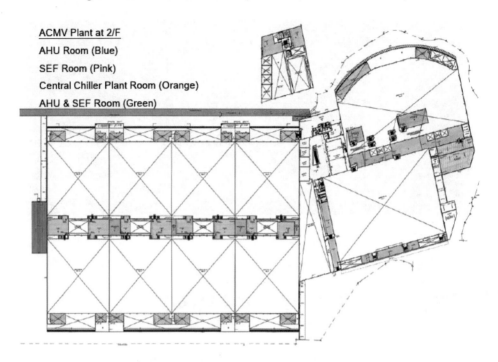

ACMV Plant at 2/F
AHU Room (Blue)
SEF Room (Pink)
Central Chiller Plant Room (Orange)
AHU & SEF Room (Green)

pre-treated fresh air supply are designed to serve offices, small meeting rooms, corridors and other cellular rooms for providing independent on/off control and flexible zoning arrangement.

3.2 Electrical Services System

In view of the extensive site coverage area of the AsiaWorld-Expo complex, it is realised that a centralised electrical transformer room is not a cost-effective option. Instead, eight transformer rooms are located around the complex to minimise the volt-drop and cable size. A total of twenty-two transformers with each of 1,500KVA, have been installed to serve the whole complex.

The emergency power supply system consists of emergency diesel generators, un-interruptible power supply (UPS) and battery & chargers. Four gensets with a total capacity of around 6MVA are installed to serve the essential services and fire services installation of the complex. In addition, 220 kVA UPS batteries have been installed to serve as emergency lighting for halls and as backup power of IT and telecom systems.

The design lighting levels of the complex are mainly based on the recommendations of the Code for Interior Lighting published by CIBSE. The halls are designed on an average illuminance of 400 lux switchable to 200 lux. In the multi-purpose hall, dimmable tungsten to 300 lux for events plus dimmable gangway lighting for maintaining at low lux level during performance have been provided for safe evacuation of occupants if necessary.

3.3 Fire Services System

The fire protection installation for the AsiaWorld-Expo complex comprises automatic sprinkler, fire hydrant/hose reel, fire detection and alarm, street fire hydrant, audio and visual advisory, visual fire alarm systems and potable fire extinguishers at the strategic locations. In general, the provisions of the systems comply with the relevant standards and requirements of the Hong Kong Fire Services Department and other recognised international standards. Centralised fire services water tanks and pumps are installed to serve the whole complex.

Besides, application of "Performance-Based Fire Engineering Approach" for huge compartmentation, life safety evacuation, strategic fire alarm zoning and cost-effective fire rating provision have been employed.

3.4 Plumbing, Drainage and Gas Supply Systems

The plumbing and drainage installations include cold water, flush water, cleansing water, planter watering, kitchen fresh water, hot water, hydro-vent water supply, foul water and storm water drainage systems. The system provisions comply with the relevant standards and recommendations of the Water Supplies Department and Buildings Department.

Metered water is directly fed to water supply outlets for the cold water systems. Hot water is provided for showers by instantaneous electric water heaters to minimise the pipe runs as the showers are located at different remote locations.

In view of the extensive roof areas of the AsiaWorld-Expo complex, siphonic drainage system is used to minimise the pipe size and falls. Town gas is supplied to the kitchens at levels 1 and 2 for catering.

3.5 *Extra Low Voltage (ELV) Systems*

Comprehensive ELV systems have been provided to facilitate the effective day-to-day operation and exhibition needs of the centre. These include advanced communication network, integrated security control system, public address system, communal aerial broadcast distribution SMATV system and building management system.

In implementing the system provisions, a hall-by-hall basis has been adopted. With the O&M arrangement, a security office is assigned to oversee the day-to-day operation of the whole complex.

In provision for the potential future development of the complex, all ELV systems have been designed such that there will be minimum design changes if expansion is needed.

4 Advanced Building Services Systems Description

The following sections elaborate the advanced building services systems provisions for the AsiaWorld-Expo complex.

4.1 *Innovative MEP Utilities Tunnel Supply Network Arrangement*

As a world-class exhibition centre designed to accommodate various demands of exhibitors, power, water, drainage, compressed air, etc. cannot be exhausted. An effective supply network planning with short set up time will lead to a successful exhibition centre. Various methods have been adopted for the utilities arrangement around the world. These include underfloor utilities pits, underfloor utilities trenches, underfloor utilities tunnels, utilities hung at ceiling high levels and dropped down to the exhibition booths (American style), etc.

According to the decision of the owner, compressed air is not provided. Instead, powers cables, cold water pipes and drainage pipes are provided and laid inside the underground services tunnel. Cables are installed at one side with water pipes laid at the

Figure 10.7 Utilities tunnel arrangement in AsiaWorld-Expo

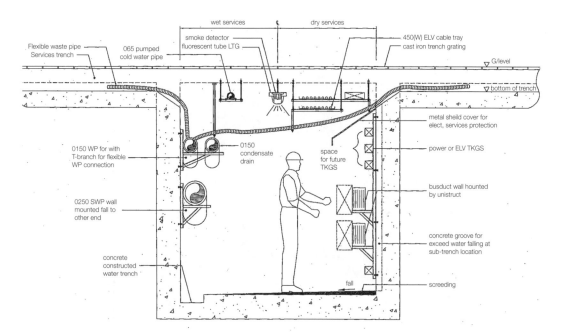

opposite side to avoid operators walking nearby getting accidental electric shock. Figure 10.7 shows the utilities tunnel arrangement in the AsiaWorld-Expo complex.

4.2 Ice Storage System with 19,000 ton-hour Ice Storage Capacity

Partial thermal storage system is adopted in the AsiaWorld-Expo complex for reducing the chiller plant installed capacity. Priority is given to chillers which provide cooling energy in day time. When the demand is higher than chiller capacity, the ice storage system provides the short fall in energy. Figure 10.8 illustrates the operating principle of the thermal storage system.

The ice storage system consists of three vertical ice tanks of 468m³ (5m dia. x 25 m high) housing the ice balls, as shown in Figure 10.9. The system produces ice and stores total 19,000RTH at night for daytime use. The system reduces total chiller capacity from 6,600 RT to 5,044 RT. There are totally 1,400 m³ storage tanks housing for 500,000 ice

Figure 10.8 Operating principle of thermal storage system

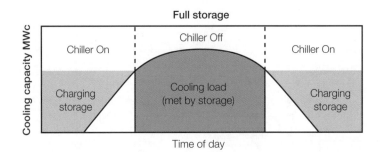

Full storage

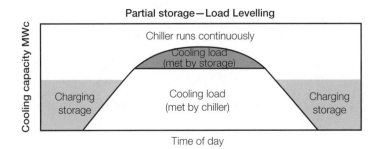

Partial storage—Load Levelling

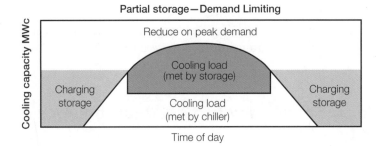

Partial storage—Demand Limiting

balls. The ice balls are filled with phase change material (PCM) inside. Once filled with PCM, the nodule plugs are sealed by ultrasonic to ensure perfect water-tightness.

The operating cycle of the ice storage system comprises 2 modes—charging and discharging modes with the phase change temperature for the ice ball at 0°C. During the nighttime/low load period, the chillers are used for charging up the ice balls by crystallisation of the salts contained within the nodules (charging mode). This takes place when the temperature of the heat transfer fluid passing through the tank is lower

Figure 10.9 Ice storage tanks in AsiaWorld-Expo

than the phase change temperature of the salts. During the daytime/high load period, the stored energy is released by fusion of the salts contained in the ice balls (discharging mode). This takes place when the temperature of the heat transfer fluid passing through the store is higher than the phase change temperature of the salts.

The thermal storage system does not only reduce the chiller plant installed capacity, but also brings the advantage of balancing power energy for daytime and nighttime. The effect will be more significant in Autumn and Winter.

4.3 New Air Ductwork System with Textile Air Distribution System

In general, ACMV installation uses sheet-metal ducting for air distribution. The principle is to distribute the required supply airflow in the occupied volume according to the internal and external thermal demand.

Air distribution with high induction textile ducts is based on a different concept. Supply air is used as a dynamic force to put ambient air in motion, with the aim of achieving high mixing rate. The technology discharges a big quantity of micro-jets through perforations made all along the duct, with velocities within the range 9–12m/s. Each

Figure 10.10 Textile duct installation inside exhibition halls

micro-jet of air is able to induce around it a secondary quantity of air by Venturi principle. The induced airflow grows and increases with air velocity.

In cooling mode, the hot air in the room will mix quickly with cool discharged air because of the direct induction all along the duct and also hot air ascension created by negative pressure beneath the duct. Air circulation is consequently controlled in the occupied volume where air discharge is evenly distributed.

The textile ductwork system provides the advantage of short installation time. It is washable and can be taken off for cleaning to improve indoor air quality of HVAC system. Other benefits for using the textile duct include its light-weight and that there is no need to provide thermal insulation around the ducts.

4.4 Extensive Usage of Siphonic System for Storm Water Drainage

The total roof area of the AsiaWorld-Expo complex is around 50,000m². Siphonic drainage system is used to minimise the pipe size and falls. Around 300 dia.90mm rain water outlets have been installed to cater for the extensive storm water drainage. The pros and cons of the conventional and siphonic roof drainage systems are summarised for reference as follows.

Table 10.1 The pros and cons of the conventional and siphonic roof drainage systems

System	Advantage	Disadvantage
Conventional System	• Suitable for all roof area • Simple drainage system	• Larger size of down pipes and rainwater outlets • Larger ceiling void space is required for pipe run due to provision of pipe slope
Siphonic System	• Smaller size of down pipes and rainwater outlets • Less space requirement for pipes and installation • No gradient is required for horizontal pipe run with greater flexibility in pipe installation • Self-cleaning in the pipe system through the full bore filling and high speed of the water in the pipe • Ease for installation works	• Not suitable for small roof area • Supporting substantiations required for statutory approval • High workmanship for pipeworks installation is required

4.5 Strategic Plant Rooms Location of Electrical Power Supply

11 kV voltage (HV) distribution network in the AsiaWorld-Expo complex is provided by the power company. The transformer substations are strategically located around the site in order to eliminate long HV cable running and HV equipment delivery inside the building. Twenty-two 1500 kVA transformers with total capacity of 33,000 kVA are distributed in eight transformer rooms, providing power supply to the AsiaWorld-Expo complex.

Except for the switchboard of the ACMV central chiller plant, each switchboard in the complex is fed by three transformers and three incoming breakers which will be electrically and mechanically interlocked to a full rating bus coupler. Castell key mechanical interlock is provided for the incomers and bus coupler to prevent any parallel operation of any two incoming transformers.

Figure 10.11 Electrical plant rooms at G/F

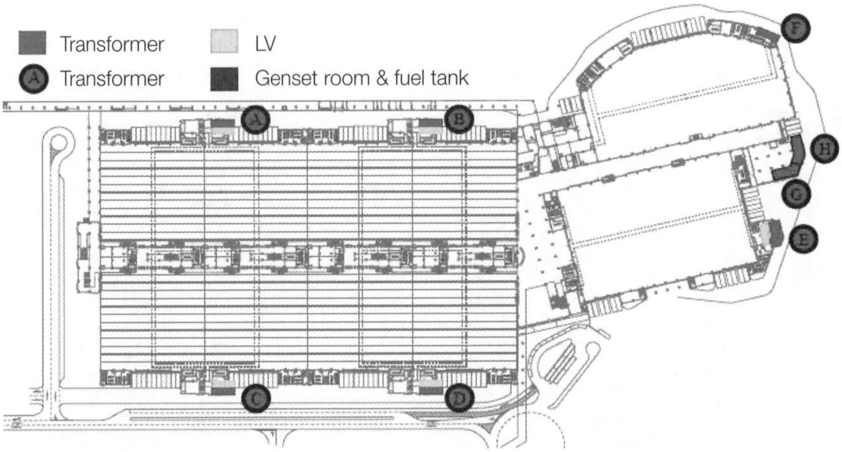

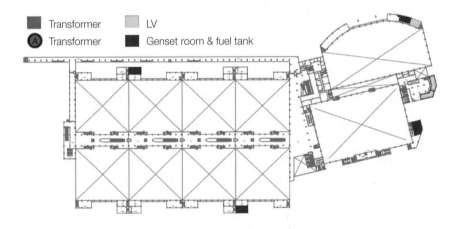

Figure 10.12 Electrical plant rooms at 1/F

Each switchboard is allocated to serve an approximate 11,000m² exhibition hall and other facilities area. The electrical loading of the exhibition hall and facilities area is evenly shared by each transformer. Figures 10.11 and 10.12 highlight the arrangement of the transformer rooms and switch rooms within the AsiaWorld-Expo complex.

4.6 Performance-Based Smoke Extraction and Make Up Air Systems Provision

Fire engineering approach for the smoke control provision has been adopted. Both dynamic smoke extraction and natural venting techniques are employed for life safety/ reliability/cost effectiveness concerns. Dynamic smoke extraction system is provided for the exhibition halls and some of the concourses which do not have sufficient roof/facade openings. For most of the main concourses, natural smoke venting is used.

Smoke extraction rates ranging from 35 to 105m³/s are provided in consideration of the fire sizes and designed smoke free height. Make-up air is provided by means of natural air intake at low level from the facade of the halls through the vehicular access doors. The air intake opening will be around 1.1m high with the bottom about 2.8m above floor level. All make-up air fans will be inter-locked with corresponding smoke extraction fans to ensure that the make-up fans will not operate alone due to any malfunction of the smoke extraction fans. Obviously, it can ensure that the space be kept at a negative pressure under smoke extraction mode. Actuation time of the system is also analysed for consideration of occupant evacuation.

Figure 10.13 Smoke extraction/natural venting provisions at 1/F

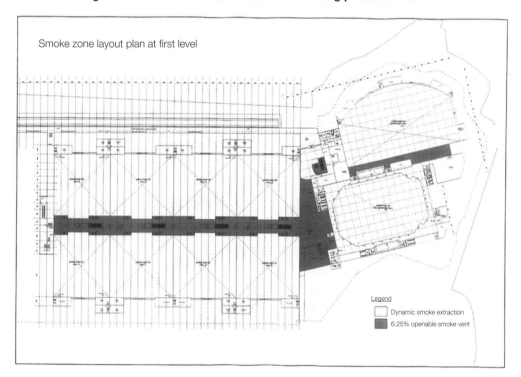

Figure 10.14 Smoke extraction/natural venting provisions at 2/F

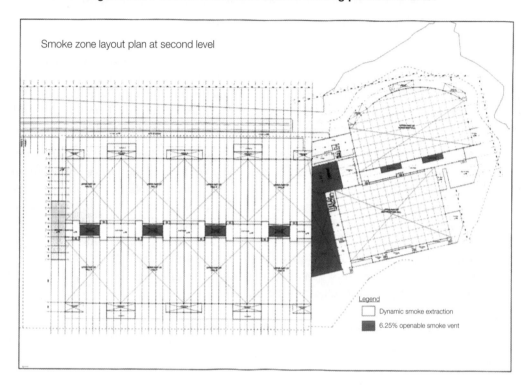

Each typical hall module of about 5,600m² is divided into three smoke zones by fixed smoke barriers inside the steel trusses. Each smoke zone is provided with dynamic smoke extraction system. Figures 10.13 and 10.14 show the smoke extraction/natural venting provisions for the AsiaWorld-Expo complex.

4.7 Application of Fire Engineering Approach for Huge Compartmentation, Life Safety Evacuation, Strategic Fire Alarm Zoning, Cost-effective Fire Rating Provisions

Owing to the functional necessity, one special feature of all the exhibition halls is their large compartment volume exceeding Building Department's prescriptive requirement of 28,000m³. If it follows the prescriptive requirement, substantial protected corridors and lobbies have to be added in order to satisfy the MOE/MOA requirements. However, this cannot satisfy the operational requirements of an exhibition hall.

Various evacuation scenarios have been carried out with different fire locations analysed. The evacuation routes, width of exits and the required evacuation time are examined and presented to the statutory authorities for approval. Besides phased evacuation, total evacuation in case of suspected bomb case is also considered.

4.8 Reliable Power Supply and Efficient Setup for the Exhibitions and Metering

Power supplies to exhibition stands, temporary catering and presentation areas in each exhibition hall are fed from a dual plug-in busbar system in a walkable underfloor services tunnel beneath the hall which allows certain supplies to be maintained on a 24-hour basis. Temporary flexible cabling will be installed to predetermined locations via empty trench and terminated in switch fuses or plug in connector as required. Either the exhibition stand electrical contractor or event promoter's personnel can install further wiring and distribution system. All temporary installations are tested for compliance with regulations before energising.

All permanently installed sub-circuits and final sub-circuits are wired in conduit/trunking system and are protected by miniature circuit breaker. Residual current-operated circuit breakers are allowed for the general power socket outlet.

Summation metering scheme is provided to maximise the benefit of the employer. A summation metering panel would be provided by the CLP in each LV switchroom for gathering data and calculation of tariff. To facilitate the energy management of the building, digital power analysers are installed for various circuits such as chiller plant, lift installation, etc. These meters are able to interface with BMS for monitoring and recording.

4.9 Green Engineering Provisions and Other Energy Saving Measures

Energy saving is the prime objective in the design of building services systems. A number of well-proven energy saving measures have been provided. It is envisaged that the exhibition centre will be characterised by largely fluctuated cooling load. In other words, this offers a large energy saving potential if proper systems, equipment and controls are provided. The salient items particularly pinpointing to reduce energy consumption in building are:

- Ice storage system
- Optimisation control for chiller sequencing
- Two-speed motor for AHUs serving exhibition/multi-purpose halls
- Free cooling mode control for AHUs serving exhibition/multi-purpose halls
- Flexible provision of mechanical ventilation at expo halls for set-up/break-down before and after the exhibitions
- Fresh air supply demand control by CO_2 level monitoring
- T5 fluorescent tubes completed with electronic ballasts to serve offices and conference rooms
- Building management system for setback control of air-conditioning and lighting on/off control in offices and conference rooms
- Power factor improvement
- Summation meterings for application of large power tariff
- Variable speed variable frequency lift motor drives

5 Conclusion

Major advanced building services systems provision in the AsiaWorld-Expo are briefly described in this chapter. In summary, these include innovative MEP utilities tunnel supply network arrangement, ice storage system by means of ice ball technology with 19,000 ton-hour ice storage capacity, new air ductwork system with textile air distribution system, extensive usage of siphonic system for storm water drainage, strategic plant rooms locations of electrical power supply, performance-based smoke extraction and make up air systems provision, application of fire engineering approach for huge compartmentation, life safety evacuation, strategic fire alarm zoning, cost-effective fire rating provisions, reliable power supply and efficient setup for the exhibitors and metering, green engineering provisions and other energy saving measures.

Figure 10.15 Exhibition held in AsiaWorld-Expo

Figure 10.16 Conference held in AsiaWorld-Expo

11

Application of Building Physics Technology on Sustainable Building Design of Hong Kong and Mainland China

Sustainable building design has an important role in today's world; it is far more than a fashionable term in a company social responsibility report, instead, it is a very real requirement. A sustainable development meets the needs of the present generation without compromising the ability of future generations to meet their own needs.

In this chapter we illustrate the principles of sustainable design through two projects that have received local and international Green Building Awards: the Hong Kong Disneyland MTR Sunny Bay Station and Beijing 2008 Olympic National Stadium ("Bird's Nest"). To demonstrate how building physics techniques impact the design process, we describe how the application of Computational Fluid Dynamics (CFD) and Dynamic Thermal Simulation (DTS) are used in assessing the thermal comfort environment of semi-opened spaces.

Kwok-On YEUNG

Chief Operating Officer (East Asia) & Director
Ove Arup & Partners Hong Kong Ltd.

Raymond YAU

Director, Ove Arup & Partners Hong Kong Ltd.

Rumin YIN

Associate, Ove Arup & Partners Hong Kong Ltd.

1 Introduction

To designers, planners, architects and engineers, sustainability is an essential element of their work. Unfortunately, there still exists a wide gap between innovative sustainable concepts and practical construction. "Building physics" is the bridge across this gap.

"Building Physics" is a sophisticated methodology that includes analytical methods and technologies that quantitatively evaluates elements related to building sustainability. Typical issues considered are thermal comfort, energy, water, light, acoustic and materials. With the assistance of these tools, a comprehensive range of sustainable concepts can be quantified with realistic performance parameters that are appropriate for each specific application. In this chapter, we highlight two of these building physics tools, Computational Fluid dynamics and Dynamic Thermal Simulation, and show how they are utilised in the sustainable design process.

1.1 Dynamic Thermal Modelling

Dynamic thermal Modelling (DTM) is a computational method for calculating the unsteady heat flow within the building fabric. Its objective is to provide realistic prediction on the thermal performance of a building, based on its composition and layout. Internal radiation exchange between surfaces is carried out using a radiosity method: the long-wave radiant heat flow (due to surface temperature differences) is handled separately from short-wave radiation (from the sun and lights). It is assumed that all reflections are non-specula. The short-wave radiant gain to the space is therefore the sum of the direct radiation on that surface plus that reflected (from an infinite number of reflections) from all other surfaces. Solar penetrations are calculated using standard optical theory, and distributed over the domain according to the relative positions of sun, surfaces and windows.

Based on annual local weather conditions and building material properties, the analysis can predict the annual real time conditions within the buildings in question. We monitor parameters such as variation of air temperature, air speed, relative humidity, pressure, and ventilation rate, solar transmittance and surface temperature. The information gained from this analysis helps determine performance requirements of a design and the selection of building fabric materials and other building features.

1.2 Computational Fluid Dynamic

Computational Fluid Dynamic or CFD is the analysis of systems involving fluid flow, heat transfer and associated phenomena by means of computer simulation. This technique has the ability to predict the thermal performance of a built environment. Unlike the dynamic thermal model which divides the flow field into several limited zones, CFD analysis relies on a field model that discretises the entire air flow domain into small elements. The predicted airflow is calculated by enforcing the principles of conservation in mass, momentum and energy. The governing equations and basic theory have been discussed in detail in published literature and will not be repeated here.

In a CFD simulation, the space is divided into cells; information within each cell such as air velocity, pressure, air temperature, turbulence intensity etc. are calculated. Useful information can be gained from analysing the "big picture" of building internal or external environment, and also the characteristic properties at specified location.

1.3 Outdoor Thermal Comfort

A sustainable design does not only provide energy efficient or energy conservation solutions, but should also achieve a high level of comfort within the building environment. Thermal comfort is a subjective measure and is influenced by a number of factors including air temperature, air humidity and air movement etc. A large amount of research has been performed on the study of thermal environment and human comfort. These studies aim to define acceptable indoor thermal conditions. However, there has been relatively little work done on open and semi-open spaces.

To extend the idea of thermal comfort indexes to semi-open spaces, it is essential to recognise the basic differences between indoor and semi-open spaces. While indoor environments tend to have relatively steady and controllable thermal, radiative and convective conditions (by building and mechanical services), the semi-open environment is defined by large daily/seasonal variations. Designers have far less control over the microclimatic parameters.

Thermal comfort for semi-open space is not only influenced by physiological response to the highly variable microclimatic parameters but also by psychological and cultural adaptation. It is indeed a complex issue and only a few attempts have been made to understand how the thermal environment affects users of these spaces. Due to the lack of in-depth studies, the approach for studying indoor thermal comfort is usually adopted or modified for use in a semi-open space. In this chapter, we will discuss some commonly adopted approaches:

- Fanger's Comfort Equation
- Physiological Equivalent Temperature (PET)
- The Research of Tanabe
- Givoni's thermal comfort index

1.3.1 Fanger's comfort equation

Fanger's assumption (Fanger, 1972) was that comfort can be derived from a human heat balance equation. The "predicted mean vote" or PMV is a measure of the subjective thermal sensation of a large population of people exposed to a certain environment; it is measured over a seven-point scale.[1] The PMV values reflect the mean value of the thermal satisfaction votes of a large group of people exposed to the same environment.

- +3   Hot
- +2   Warm
- +1   Slightly warm
- Neutral
- -1   Slightly cool
- -2   Cool
- -3   Cold

The PPD index represents the percentage of people dissatisfied with the thermal conditions when exposed to a particular environment. Based on empirical analysis over a large population scale, both the PMV and PPD can be calculated from prescribed formulae that are based on parameters such as relative humidity, air velocity and air temperature etc. Fanger's index is commonly used in the analysis of air-conditioned or heated spaces.

Case studies for Fanger's Comfort Equation were conducted in climatic chambers with controlled internal environment. While this is adequate in simulating an indoor environment, the subjective comfort level can be substantially different for an outdoor or semi-open environment with similar conditions. Direct application of Fanger's Comfort Equation is not suitable for outdoor conditions.

1.3.2 Physiological equivalent temperature (PET)

The physiological equivalent temperature, PET, is a thermal index derived from the human energy balance. It takes into account the dominant meteorological parameters influencing the human energy balance; this includes air temperature, vapour pressure, wind velocity and mean radiant temperature of the surroundings. From everyday experience, it is obvious that the effects of solar radiation play a significant role in determining the thermal comfort of an outdoor space. The PET methodology does not explicitly include the effects of solar radiation, and is thus only suitable for indoor thermal comfort assessment. The PET methodology is not recommended for outdoor and semi-outdoor thermal comfort assessment.

1.3.3 Bioclimatic chart

The bio-climatic chart[2] shows the relationship of the four major climate variables that determine human comfort. By plotting temperature and relative humidity, one can determine if the resulting condition is comfortable (within the comfort zone). The chart has relative humidity as the abscissa and temperature as the ordinate. The boundaries of a comfort zone demarcated on the Bio-climatic chart can be extended to hot-humid conditions, but this presentation method is complex. Furthermore, wind and shading effects cannot be included at the same time in a Bioclimatic chart.

1.3.4 Research of Tanabe

A research conducted by Tanabe et al (1988) modified the findings of Fanger's Comfort

Figure 11.1 Tanabe's mean thermal sensation votes as a function of the PMV

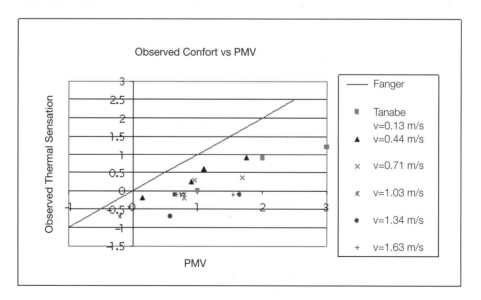

Equation and explicitly included the effect of acclimatisation and air speed. The research of Tanabe considers the thermal sensations of the population in hot-humid climates areas such as Hong Kong. The results are illustrated in Figure 11.1. The data indicates influence on thermal sensations due to air movements surrounding the occupants. Unfortunately, such studies cannot give a universal criterion for outdoor thermal comfort; the effect of solar radiation is also not explicitly included.

1.3.5   Givoni's thermal sensation index

Givoni's thermal sensation index (2003) is a specially designed index system that predicts the thermal comfort condition of an outdoor space based on the experimental findings from local data in Japan. The analysis considers the effects of the ambient temperature, relative humidity, wind speed, solar radiation and the protection from solar radiation (e.g. shading elements) and the local wind speed (e.g. wind breaks) on thermal sensation. Figure 11.2 illustrates the experimental results of this study. In a similar manner to the Fanger's comfort index, Givoni's index ranges from 1 to 7, representing the thermal comfort conditions from very cold to very hot.

Givoni's thermal sensation index considers all major environmental elements that affect outdoor thermal comfort levels. It includes a comprehensive range of elements such as air temperature, humidity, wind speed, solar radiation and surface temperature. Amongst the various methods described here, so far Givoni's approach is the most suitable methodology for outdoor thermal comfort assessment.

Figure 11.2 Experimental results of Givoni's thermal sensation index analysis

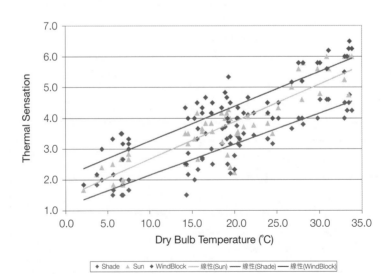

2 Case Study—Beijing 2008 Olympic National Stadium

The Beijing National Stadium is the main stadium of the 2008 Beijing Olympics. It has a net seating capacity of 100,000 spectators in the "Olympic mode" and 80,000 in the post-Olympic mode.

This winning design for Beijing National Stadium is dubbed the "bird's nest" because of its innovative grid formation. The roof is integrated with a retractable canopy that can be extended to cover the whole stadium, transforming it from a semi-opened area to fully covered area.

A single layer transparent, water proof ETFE membrane is used as the outer roof. On the underside of the steel truss is an acoustic ceiling of PTFE membrane materials. The two layers together will provide a high level of light transmission. The roof also acts as a giant umbrella that shelters the stadium from the sun, and rain during the games but will also allow a natural exchange between ambient and internal air.

The National Stadium is one of the icons of the 2008 "Green Olympics", it includes green features such as a ground source heat pump system, variable control system, free cooling system and environmental friendly construction materials etc.

The thermal conditions are critical, especially when the stadium is in "Olympic mode", during which up to 100,000 spectators can be housed within the structure. How do we determine the comfort level of a stadium? How do we know if it is good enough? Are active ventilation systems such as mechanical fan or even air conditioning systems necessary? It is essential to understand the performance of natural ventilation inside the stadium by means of the proposed ventilation openings at low and high level before we can answer these questions.

With this objective in mind, the internal thermal environment was studied under many different conditions. To determine the thermal comfort performance of the stadium, the resulting temperature within the stadium, especially at the spectator area, was assessed. The most uncomfortable conditions are predicted at the high tiers (black-white in Figure 11.3).

To better understand the thermal comfort level of the spectator area, the solar heat gain, surface temperature must be evaluated first. This part of the study utilises DTM for the building model. The analysis predicts the temperature and ventilation rate of the spectator area by using heat and mass balance equations and buoyancy factors for heat dissipation inside the stadium. The program takes into account the thermal properties of the stadium structure, building shape, site orientation and location, area and orientation

Figure 11.3 Competition design of Beijing Olympic National Stadium

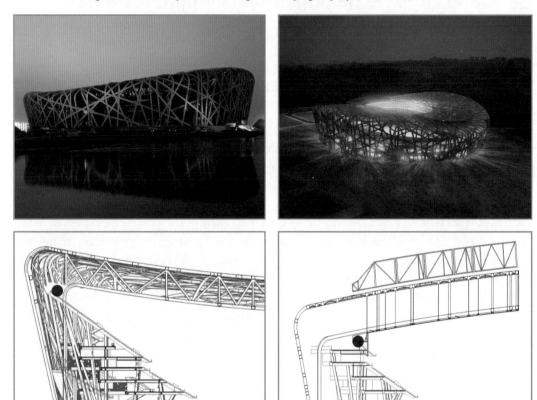

of openings, external wind data, air temperature, solar radiation and internal heat load depending on occupancy. It can generate hourly values of surface temperature, air change rate, dry bulb temperature and resultant temperature of the studied space.

Figure 11.4 shows the temperature variation for the wall surfaces of the stadium in July and August. The surface temperatures of the void steel structure and roof steel structure could increase to 50°C and 55°C respectively.

Figure 11.5 shows the surface temperatures of the roof and steel members in a typical day in August (i.e. to simulate conditions during the Olympic Games). For the selected day, the maximum temperature of acoustic ceiling and roof cladding could increase to 38°C during daytime. The roof steel member could be even higher: up to 47°C due to strong solar radiation effect and heat absorption property of steel. The highest ambient air temperature is 31°C during daytime, and 28°C in the evening at 7:00 p.m.—the time for the opening ceremony of the Olympic games.

Figure 11.4 Variation of wall surface temperature and outdoor temperature from July to August

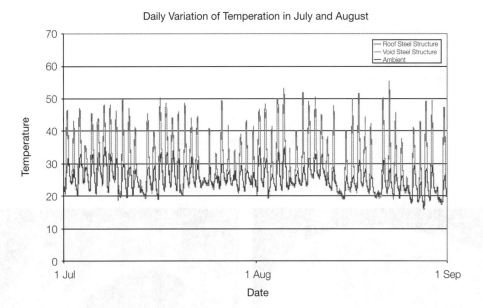

Daily Variation of Temperation in July and August

Figure 11.5 Variation of surface temperature and outdoor temperature in one typical day in August

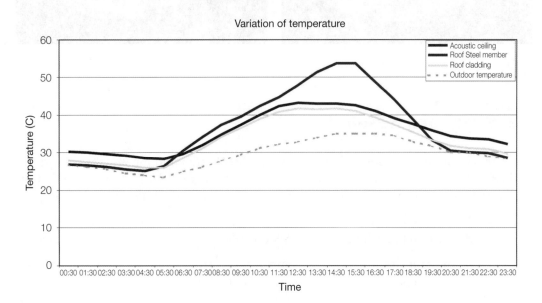

Variation of temperature

287

2.1 CFD Model

A 3-dimensional model of the entire stadium was constructed for this study; it includes the major architectural layout: the curved double layer ceilings (roof and acoustic ceiling), upper, mid and low spectator areas and openings at the level 1 and the giant screen/scoreboard. The CFD simulation targeted the worst case scenario—peak spectator load during warmest summer conditions in August. Figure 11.6 is the simulation model and grid.

Figure 11.6 External and internal view of the CFD model

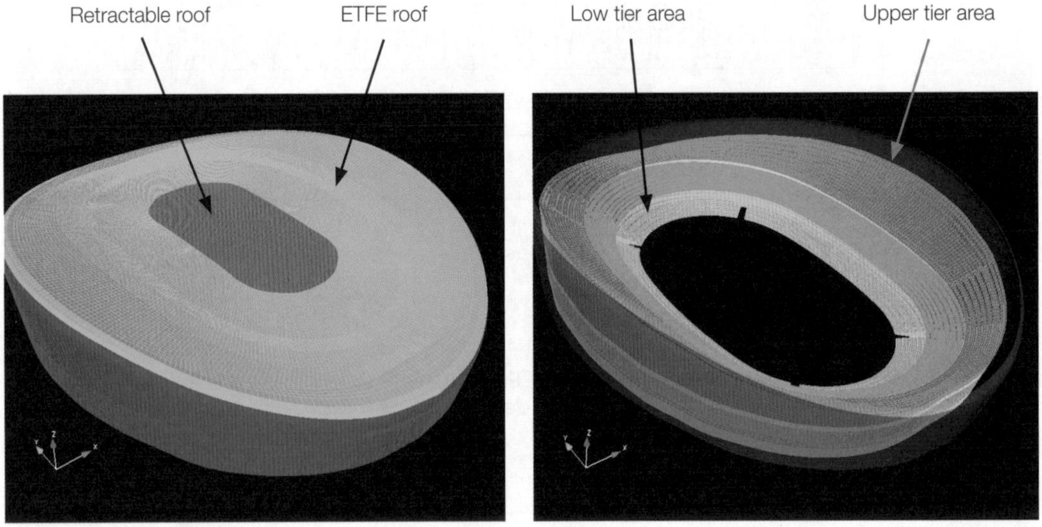

Retractable roof   ETFE roof   Low tier area   Upper tier area

2.2 Simulation Result

The temperature distribution, relative humidity distribution and air velocity vectors are evaluated and the results are shown in the following figures. With the retractable roof closed, the air movement within the stadium is a function of the buoyancy effect only:

- The result shows that the air temperature for low tier area is between 29°C and 30°C with relative humidity of 75% to 80%.

- For the mid tier area, the air temperature varied from 30°C to approximate 32°C with relative humidity around 70% to 80%.

- For the high tier area, the temperature ranges from 31°C to approximate 37°C with relative humidity around 60% to 73%.

- In general, higher temperature and lower relative humidity are found at higher location.

- With limited open space at the roof, the air flow movement generated by buoyancy effect is limited; air speed is less than 1 m/s in most areas.

Figure 11.7 Temperature distribution

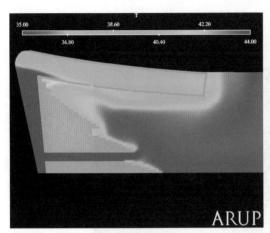
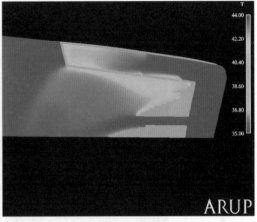

E- W section view   N- S section view

Figure 11.8 Velocity vectors

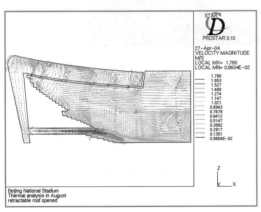
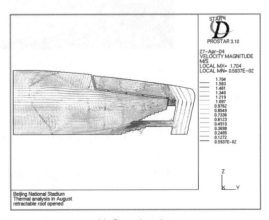

E- W section view   N- S section view

2.3 Thermal Comfort Analysis

Based on the results from the Dynamic Thermal and Computational Fluid Dynamic simulations, the thermal comfort conditions at the spectator area of the stadium are calculated.

As shown in Figure 11.9, the thermal sensation index (see section 1.3.5) ranges from 4.25 to 4.75 in the low tier, 4.5 to 5.0 in the mid tier and 4.5 to 5.5 in the high tier. The spectators are expected to feel hot at these high tier areas if no additional cooling strategies are used to generate air movements.

Figure 11.9 Givoni's thermal sensation index at the spectators' area

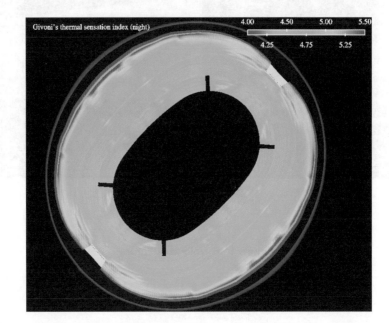

2.4 Design Optimisation

Unfortunately, it is neither feasible nor cost-effective to provide mechanical ventilation system for the identified "hot zones". To achieve the ideal of a "green Olympic", the following moderate strategies have been incorporated in the final design:

- Reduce the seating capacity from 100,000 to 91,000 during the Olympic Game period, so as to reduce the internal load.

- Increase the distance from the highest seat to the false ceiling from 2 m to 8 m, such that the occupants are seated below the stratified hot air layer beneath the roof

- Omission of retractable roof and enlargement of the roof opening area enhances the natural ventilation effect and also has the added benefit of reducing structural materials.

- Reduce the coverage area of outer transparent ETFE membrane layer in order to enlarge the open space for natural ventilation at the side

2.5 Result of Optimised Design

The CFD simulation is also applied to evaluate the thermal performance of the optimised design. The model is depicted in Figure 11.10.

Under the same outdoor condition as described in the previous section, the modified design shows significant improvement in thermal performance when compared to the previous design:

- At the low tier area, the temperature ranges between 29°C and 30°C.

- At the mid tier area, the temperature ranges from 30°C to 31°C.

Figure 11.10 CFD model for the latest design

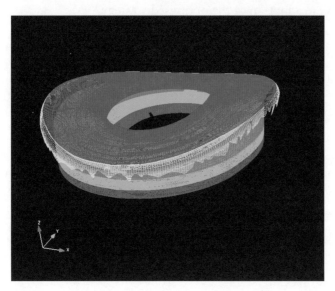

Figure 11.11 Givoni's thermal sensation index for the optimised design

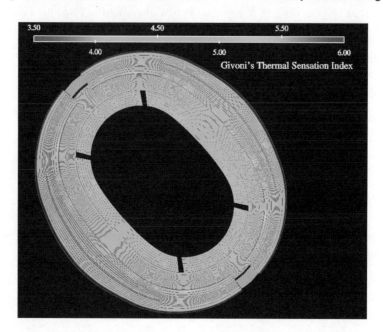

- At the high tier area, the temperature ranges from 31°C to 33°C.

- The maximum temperature in the compound decreased from 37°C to 33°C through the adoption of the optimised strategy.

- Compared to the ambient temperature of 28°C, this represents an improvement of 45% in terms of thermal environment.

- The evaluation of thermal sensation index also shows that during night-time operation, with the exception of some localised hot zones, the thermal sensation index in most areas varied from 4.0 to approximate 5.0, this is within the defined comfort zone.

With the aid of "Building Physics" analysis, the built environment can be improved through design optimisation, even without additional active thermal systems.

3 Case Study—Hong Kong MTRC Disneyland Resort Sunny Bay Station

Sunny Bay Station was opened in June 2005; it is an interchange station between the MTR Tung Chung Line and the Disneyland Resort Line. The station is also the starting point of the "magical journey" to the Hong Kong Disneyland at Penny's Bay.

Sunny Bay is designed to be a sustainable and green building. The gentle curving fabric roof, made of Polytetrafluoroethylene (PTFE) membrane, is on a lightweight steel roof structure. The structure creates a pleasant semi-opened environment sheltering occupants from intense solar radiation, wind and rain. Most importantly, it also significantly reduces energy consumption through appropriate use of natural ventilation and daylight in place of fully air-conditioned stations. Moreover, the supplementary water mist system operated under extreme hot weather condition can provide cool and pleasant environment with only 1/3 of energy consumption compared to typical systems.

Figure 11.12 Sunny Bay station

3.1 Sustainable Consideration

The design of Sunny Bay station considers the comprehensive range of sustainable items to enhance its environmental performance. The positive elements are identified as follows:

- Use of energy efficient technology
- Optimal ventilation and daylighting design

- Use of green construction materials
- Maximised utilisation of the land
- Enhanced international competitiveness and attractiveness
- Financially viable
- Positive contribution to local economy by enhancing tourist attraction
- Easy access to public transport interchange
- Environmentally friendly transport

3.2 Thermal Comfort Analysis

Sunny Bay Station is a naturally ventilated space. Sufficient natural lighting is also provided over most of the daylight hours. Thermal comfort is an important issue and needs to be considered carefully at the beginning of the design process.

Figure 11.13 Sustainable items in the Sunny Bay Station

The design methodology adopted here is similar to that used for the Beijing National Stadium. Both DTM and CFD simulations are applied for the thermal comfort analysis.

However, due to the difference in functions of the two developments, the analysis must be modified to be tailored to the nature and characteristics of the building; a cookie cutter type approach will not be appropriate here. For example, unlike the national stadium, Sunny Bay Station is operated throughout the year for at least 15 hours each day. The analysis needs to consider the performance under these conditions.

3.3 Dynamic Thermal Modelling Simulation

In order to investigate the performance of the internal environment of the station under various operation scenarios with variations on train heat load and platform occupant loads, the following scenarios were studied by DTM:

Table 11.1 Simulation scenarios

Case	Train Heat Load	Platform Occupant Load	Situations Occurred
1	Full loaded	Maximum	This is the worst case. It is representative of the peak hour when full-loaded trains dwell at the platform, large number of passengers are waiting at the station to board the train, or waiting for another train.
2	Nil	Maximum	Passengers are waiting at the station for the trains during peak hours.
3	Empty loaded	Maximum	Trains have just arrived and passengers are waiting to get onto the trains.
4	Full loaded	Nil	Passengers have boarded the trains and there are no passengers left at the station.

Scenarios 1 and 2 happen frequently during the daily operation of Sunny Bay Station, while Scenarios 3 and 4 are transient situations, as a result, scenarios 1 and 2 are adopted for CFD simulation.

Figure 11.14 shows the typical result of the resulting temperature of the occupied zone demonstrating the effects on the variations of both train loads and occupant loads.

Figure 11.14 Temperature profile

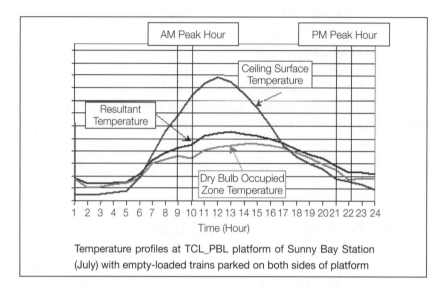

Temperature profiles at TCL_PBL platform of Sunny Bay Station
(July) with empty-loaded trains parked on both sides of platform

3.4 Computational Fluid Dynamic Simulation

The CFD simulation is also carried out for the internal environmental analysis. For
a comprehensive study of the thermal environment, the scenarios do not only cover

Figure 11.15 CFD model of Sunny Bay station

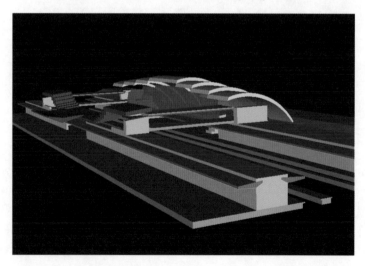

Figure 11.16 CFD result under windless condition with train parking

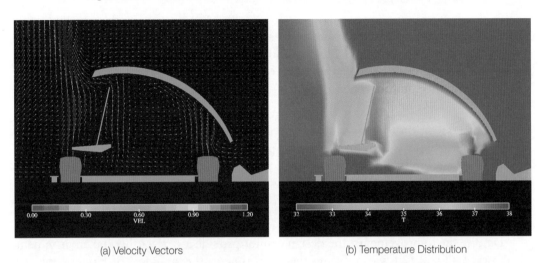

(a) Velocity Vectors   (b) Temperature Distribution

Figure 11.17 CFD result under prevailing wind condition with train parked at platform

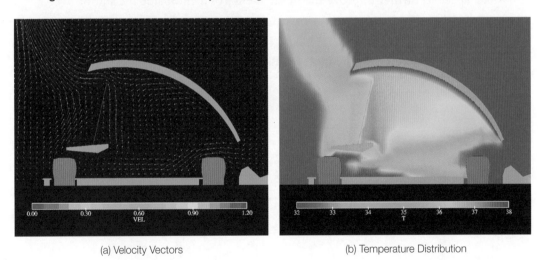

(a) Velocity Vectors   (b) Temperature Distribution

windless condition, but also those under prevailing wind condition. The above drawings show the CFD model and the typical result under different conditions.

3.5 *Thermal Sensation Index*

The thermal sensation index is assessed to evaluate the thermal performance of Sunny Bay Station. As shown in Figure 11.18, although the thermal comfort level is acceptable during the evening peak period, when the visitors are returning from Disneyland, it can still be beyond the comfort zone (too warm) during morning peak time and most of day time.

The investigation found that the opening is large enough to permit sufficient air flow. Solar heat gain is also an important parameter that needs to be controlled. The optimised design limits the sky light area at the roof while extending the shaded area at the north facade. This provides good shading effect as well as daylighting. The optimised design shows a significant improvement in thermal environment, as shown in Figure 11.19.

Despite the improvements, the design cannot provide a perfect thermal environment since users will still feel warm during the noon time in the summers of Hong Kong. To moderate this, a cost-effective and energy efficiency water mist system is designed for the semi-open space to further improve the thermal environment inside the station.

Figure 11.18 Thermal sensation index for original design

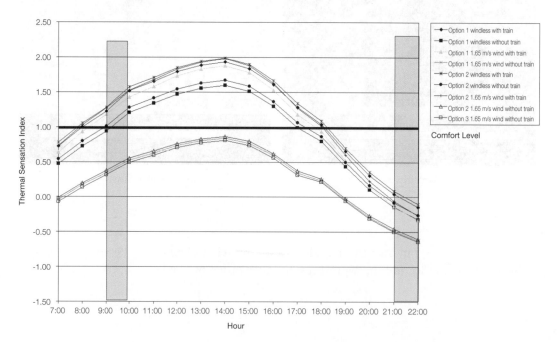

Figure 11.19 Thermal sensation index for optimised design

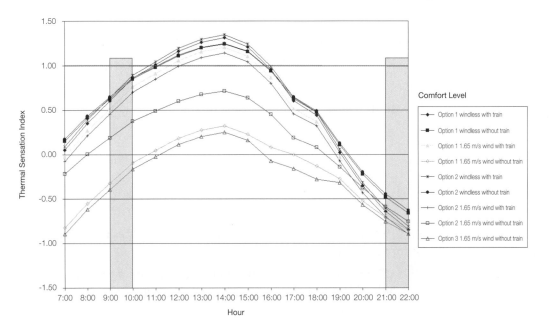

Comfort Level

- ◆ Option 1 windless with train
- ■ Option 1 windless without train
- ▲ Option 1 1.65 m/s wind with train
- ◇ Option 1 1.65 m/s wind without train
- ✳ Option 2 windless with train
- ● Option 2 windless without train
- ＋ Option 2 1.65 m/s wind with train
- ☐ Option 2 1.65 m/s wind without train
- △ Option 3 1.65 m/s wind without train

Figure 11.20 Water mist system

4 Conclusion

Building Physics is a sophisticated science and technology, the case studies presented here focused on the thermal environment of semi-open spaces analysed with DTM and CFD. Despite the similarities in analysis method, design optimisation is very different for the two cases and is based upon different determining factors such as functions of space, operation time and local climate condition etc. Sustainable design is a holistic design approach, proper use of advanced technology is essential.

Notes

1. BS EN ISO 7730. (1995). Determination of the PMV and PPD indices and specification of the conditions for thermal comfort.

2. Givoni, B. (1998). *Climate considerations in building and urban design*. New York: Van Nostrand Reinhold.; Tanabe, S. I. (1988). *Thermal comfort requirement in Japan*. Waseda University, Tokyo, Japan.

References

De Dear, R., Fountain, M., Popovic, S., Wakins, S., Brager, G., Arens, E., et al. (1993). A field study of occupant comfort and office thermal environments in hot-humid climate. *ASHRAE RP–702.*

De Freitas, C. R. (1985). Assessment of human bioclimate based on thermal response. *International Journal of Biomterol,* Vol. 29, No. 2. pp 97–119.

Fanger, P. O. (1972). *Fanger thermal comfort.* New York: McGraw-Hill Book Company.

Givoni, B. et al. (2003). Outdoor comfort research issues. *Energy and Buildings,* Vol. 35, pp. 77–86.

M. A. Humphreys. (1976). Field studies of thermal comfort compared and applied. *BSE Vol. 44.* pp. 5–27.

Nicol, F., & Raja, I. A. (1997). Modeling temperature and human behaviour in buildings. *IBPSA News, Vol. 9, No.1.* pp. 8–10. UK: London.

Versteeg, H. K., & Malalasekera, W. (1995). *An introduction to computational fluid dynamics—The finite volume method.* Harlow England: Addison Wesley Longman Limited

Awards

The Sunny Bay Station is recognised as a successful sustainable design, it has received a number of international green building awards such as "The First Hong Kong Green Building Award 2006—Grand Award", "2007 Europe BEX sustainable Building—Runner Up Award", "The 2nd Biannual Business Week/Architectural Record China Awards—Best Green Design".